God's Canvas

AN EXPLORATION OF FAITH,

ASTRONOMY, AND CREATION

Fr. James Kurzynski STL

FOREWORD BY BR. GUY CONSOLMAGNO

God's Canvas: An Exploration of Faith, Astronomy, and Creation

Authored by Fr. James Kurzynski STL
Foreword by Br. Guy Consolmagno, S.J.
Edited by Ann Del Ponte
Prepared for Publication by Travis J. Vanden Heuvel

ISBN: 978-0-9969426-2-1

A Peregrino Press book

De Pere, Wisconsin

Peregrino Press seeks to publish material that is free of doctrinal or moral error. *God's Canvas* has been submitted to Church authorities in the Diocese of La Crosse, WI and has been granted the following:

NIHIL OBSTAT:
Rev. Justin J. Kizewski
October 12, 2016

IMPRIMATUR:
+Most Rev. William Patrick Callahan
October 17, 2016

DEDICATION

To Mr. Ron Davis, retired English teacher at Amherst High School:

In 1991, you encouraged one of your non-college bound English students to transfer into college-bound English and consider going to college. Without your encouragement, this book, and my priesthood, may never have happened.

To the Jesuits at the Vatican Observatory:

In gratitude for taking a big risk by pursuing the idea of a complete stranger--your willingness to be open to new ideas has inspired me to do the same in my life.

To those who have supported my dreams:

Thank you to everyone who has given me the space and encouragement to pursue what God has laid upon my heart; in particular, my family, friends, brother priests, former students, parishioners, Bishop William Patrick Callahan, my past bishops, and all who continue to inspire me to embrace my vocation as a diocesan priest.

CONTENTS

Foreword... 1
Br. Guy Consolmagno

Introduction.. 3
Let the pilgrimage begin!

Section One: Aren't faith and science at war with one another?

Introduction.. 9

Chapter 1.. 11
Laying the Foundation
Why do debates on faith and science almost always fail?

Chapter 2 21
Exploring our Origins
C.S. Lewis and the fight for meaning in Genesis

Chapter 3 29
Steven Hawking
Didn't Dr. Hawking disprove the need for God?

Chapter 4 37
Foundations of Creation
What is creation Ex Nihilo and Creatio Continua?

Chapter 5 41
The Catholic Church and Evolution
Isn't the Church against evolution?

Chapter 6 47
Early Language of Spiritual "Evolution"
Irenaeus of Lyons and the Opulence of God

Conclusion.. 53

Section Two: Big thoughts, big thinkers – finding common ground between faith and science

Introduction.. 55

Chapter 7 57
George Lemaitre
Father of the "Big Bang"

Chapter 8 61
Stanley Jaki, OSB
The Priest who questioned the plausibility of a Theory of Everything.

Chapter 9 65
Women of Science, Women of Faith
Reflecting on influential women of faith and science

Chapter 10.. 71
Teilhard de Chardin and Catherine Pickstock
Understanding the Language of Creation.

Chapter 11 77
G.K. Chesterton
So a pigmy goes to Mass in a multiverse and experiences a severe case of anamnesis.

Chapter 12..83
Flannery O'Connor
Faith, imagination, creation, and co-creation

Chapter 13..87
Carl Sagan
Perspective on the "Pale Blue Dot"

Chapter 14...91
Saint Bonaventure and Dr. Michio Kaku
Has String Theory proven the existence of God?

Chapter 15..99
Thomas Aquinas
Does the Catholic Church need someone to assimilate science into theology?

Conclusion...103

Section Three: Understanding the "Cosmic Liturgy"

Introduction..105

Chapter 16...107
The Hymn of Creation
Sun and Moon, Bless the Lord

Chapter 17...109
Advent
Amid creations groaning, there is hope

Chapter 18..113
Christmas
He is Born!

Chapter 19.. 117

Epiphany

How all of creation points to our source and summit

Chapter 20 121

The Transfiguration

The limits of language and the power of metaphor

Chapter 21.. 127

Easter Vigil

The beauty of this night

Chapter 22 131

Sacrament of Confirmation

Fear of the Lord – A gift of the Holy Spirit meets astronomy

Chapter 23.. 137

Measured Time and Sacred Time

So when is Easter?

Chapter 24 141

Beginnings and Ends

O death, where is your sting?

Conclusion.. 145

Section Four: For the love of astronomy – Light hearted reflections and significant events from 2015–2016

Introduction...147

Chapter 25...149
Faith and Astronomy Are a Waste Of Time!
Or... maybe not

Chapter 26 155
Full Moons
Looking for life in all the right places

Chapter 27 163
Fibonacci Numbers
What do they tell us about our world (if anything)?

Chapter 28 167
The Discovery of Gravitational Waves
Is there a "music" to the universe?

Chapter 29 171
Pluto Flyby
The dwarf planet that did not disappoint!

Chapter 30 175
Pierre Gassendi
The transit Mercury, the priest who recorded the data, and exploring where we go from here

Chapter 31 179
Whirlpools, Sunflowers, and Pinwheels
Astronomy on a lazy Sunday afternoon

Conclusion...183
And the pilgrimage continues!

FOREWORD

BROTHER GUY J. CONSOLMAGNO, S.J.

Brother Guy J. Consolmagno, S.J. is an American research astronomer with B.A. and M.A. degrees from the Massachusetts Institute of Technology, and a Ph.D. from the University of Arizona's Lunar and Planetary Laboratory, all in planetary science. He believes in the need for science and religion to work alongside one another rather than as competing ideologies. Known as "The Pope's Astronomer," Br. Guy was named by Pope Francis to be the Director of the Vatican Observatory in September 2015.

THERE ARE A LOT of books out there about science and religion. I've written a number of them myself. But there are a lot fewer *good* books on the subject.

This is one of the good ones.

I first met Fr. James over the Internet when he wrote me out of the blue with the crazy idea that the Vatican Observatory could host an annual workshop on astronomy for Catholic pastors and teachers. That crazy idea has become one of the key events in the work of the Vatican Observatory Foundation. It's made for a bit of work on our part, but it's been immensely rewarding. And when Jim and I finally got to meet, peering through a telescope at the first of those workshops at a retreat house in the desert outside Tucson, I discovered that he certainly knows his astronomy.

But since he'd made all that work for me, I figured I would rope him into writing articles for our Foundation's blog, *The Catholic Astronomer*. I was expecting that he might pass on a few lightly rewritten Sunday homilies. Instead, he immediately sunk his teeth into some of the thorniest issues in science and faith. And if that mix of metaphors makes you wince (my mouth hurts just thinking about teeth into thorns), you can imagine how I felt when I saw some of the topics he was approaching… from the literal nature of scripture, to the science and philosophy of climate change. It's so easy to get these wrong. But when I read his articles, I didn't just feel relief that "he got it right" — i.e., he agreed with me! Rather, I was taken to deeper insights into some of the very topics that I myself had dared to write books about. He knows his theology, too.

He's also a darned good writer.

I forget where the idea first came from that these columns would make a fine book. I'd like to claim it for myself, but I suspect the origin of this book is a bit more complicated than that. In any event, the wisdom of that idea rests now in the hands of you, the reader.

And if you like it … consider following more of his material (and that of other writers, as well) on our ongoing blog site, at www.vofoundation.org/blog. If you're a Catholic pastor or teacher, come join us at our annual Faith and Astronomy Workshop!

Enjoy!

Br. Guy Consolmagno
Director of the Vatican Observatory

INTRODUCTION

LET THE PILGRIMAGE BEGIN!

D ID YOU EVER HAVE a dream that not only came true, but took on a life of its own? What you are about to read is the fruit of one of my dreams.

A few years ago, I was sitting in my office at Roncalli Newman Parish in La Crosse, Wisconsin, putting together an e-mail to Br. Guy Consolmagno of the Vatican Observatory. In the e-mail, I recounted how, shortly after my ordination to the priesthood in 2003, I wrote an e-mail to the then Director of the Vatican Observatory, Fr. George Coyne. In that e-mail, I asked Fr. Coyne if the Vatican Observatory offered retreats or conferences on faith and science to the non-professional scientist.

I'm not a scientist, but a priest who has had a deep interest in science since my youth. I was one of those daydreaming kids that liked to look for animals in the clouds and lie in the backyard of my parents' central Wisconsin farm to gaze into the night sky. Although I studied a little astronomy in college, I wanted more.

When I was in seminary, I had the chance to explore questions of faith and science, but upon ordination, I still felt I needed more. This background (or lack of one) is what compelled me to write to Fr. Coyne.

Fr. Coyne's response was gracious and explained that the Vatican Observatory didn't offer such programs and worked primarily in post-doctoral astronomy, but perhaps this type of programming should be explored in the future. He encouraged me to write back "in the future" to revisit the topic.

Looking back, I have to laugh a little about his suggestion. After I received his e-mail, I was busy embracing my new life as a diocesan priest. I spent time as an associate of three parishes, seven years as a middle school/high school chaplain/teacher, and a few years as the pastor of parishes and Newman Ministries on college campuses. Needless to say, the follow-up e-mail was put on the back burner.

Ten years later, I finally got around to writing back to the Vatican Observatory. Fr. Coyne had retired, and Fr. Funes was the new director. Instead of writing to the director, I decided to follow up with Br. Guy because I had seen some of his faith and astronomy videos on YouTube and was impressed.

At the time, Br. Guy was in charge of communications and public outreach for the Vatican Observatory, making him the logical astronomer to contact. After recounting my previous correspondence, I asked again if the Vatican Observatory had any programs for non-scientists that addressed issues of faith and science.

Again, as I expected, the answer was no. However, Br. Guy opened a door of exploration by simply stating that he liked the idea and wanted to take it to Fr. Funes for consideration.

The excitement I felt was immense. I recall being both excited and nervous as I waited for Br. Guy to share Fr. Funes' response. My excitement got the best of me and I e-mailed Br. Guy again, asking what Fr. Funes thought of the idea. The response was quick and affirming: He loves the idea, and we will begin planning for the first Faith and Astronomy Workshop (FAW) in Tucson, Arizona, for 2015.

I remember reading the e-mail at least three times, almost in disbelief that a simple request for something I had hoped already existed was now about to lead to something that had never been done before by the Church and I was going to be one of its first participants! This development, in and of itself, would have been more than enough for me to feel a sense of accomplishment when it came to this dream. However, God had more surprises that I never could have imagined.

As the weeks passed, the first Faith and Astronomy Workshop finally arrived. In the run-up to the event, things started to dawn on me that gave me moments of concern. For one, the only contact I had had with Br. Guy was via e-mail. I had never met or talked with him, and now I was going to an event that he and the Vatican Observatory staff put together based on MY idea.

I started to worry, "What if it fails?" I had been to many theology workshops before, but this felt radically different given the personal connection I had made in the development of the program.

Happily, the workshop went well with some wonderful highlights, but also some predictable needs for improvement. After all, nothing like this had ever been done before so why wouldn't there be room for improvement?

I could write a chapter a day on what we did, but the summary is that Br. Guy, Fr. Gabor, Fr. Corbally, and friends of the Vatican Observatory Foundation threw us into the world of professional science and encouraged us to apply our faith background to discover the bridges between faith and astronomy on our own. This "no spoon-feeding" approach was a little awkward at first, but bore a great deal of fruit. In short, I came away feeling joy that one of my dreams had been realized.

When the first Faith and Astronomy Workshop was over, I remember sitting in the Tucson airport. I was feeling a slight twinge of sadness. The sadness stemmed from the fact that I had an idea, I pursued it, it bore fruit, but what was next? Was this the end of the journey? I remember a moment of prayer in the airport terminal when I felt God put a simple sentiment on my heart, "You don't need to do anything more. I will open doors that need to be opened in the future."

This prayer brought relief, but I never could have imagined what those doors would be. The first door turned out to be an invitation to become a contributor of the Vatican Observatory Blog titled *The Catholic Astronomer*. The second door was an invitation to attend the second Faith and Astronomy Workshop in 2016 as a presenter. Lastly, the third door is in your hands (or on your screen) right now: this book, *God's Canvas*.

After the last workshop, Br. Guy encouraged me to look through my blog posts to see if "there was a book hidden in them." I am happy to report, there was!

This collection of my posts for the *Catholic Astronomer* has been separated into four sections. In Section One, I pulled together thoughts that can give us an understanding of an authentically Catholic approach to faith and science. In particular, we will explore the nature of faith and science, arguing that when looking at both on their own terms, these two great disciplines are best approached as dialogue partners in contrast to the common myth that the two are, and must be, in conflict with one another. This section will also explore the claims of Stephen Hawking that modern science deems the idea of a Creator to be not necessary, the question of whether or not Catholics believe in evolution, how Catholics read the Book of Genesis, and reflections on a spirituality of growth and evolution found in Saint Irenaeus of Lyons.

After setting our foundations, Section Two will explore some of the most profound thinkers in faith and science. We will meet the "father" of the Big Bang Theory, Monsignor Georges Lemaitre, reflecting on how he saw something rather different than did the rest of the scientific world when looking at Einstein's Theory of Relativity. We will meet Fr. Stanley Jaki, OSB, who rightly questioned the plausibility of the Theory of Everything. We will also reflect on the role of women in the exploration of faith and science, embracing Pope Francis' call to empower women in the Church where possible. We will also explore the thoughts of figures from a broad reach of disciplines including Theilhard de Chardin, Catherine Pickstock, G.K. Chesterton, Flannery O'Connor, Carl Sagan, Saint Bonaventure, and Michio Kaku. As we view faith and science through these people, we will conclude by tackling the question, "Are we in need of a new theological assimilation of faith and science in our modern age?"

Section Three will discuss the idea of "Cosmic Liturgy." Catholicism has a long tradition of seeing our liturgical prayer life as being intimately connected with the cosmos. These reflections are a mix of theological and personal views along with connections to the liturgical seasons to demonstrate how all of creation participates in what can be called the "Hymn of Creation." This hymn affirms that all things give honor and glory to God by their very existence with humanity possessing a unique voice in this hymn, being made in God's image and likeness.

The final section is a collection of reflections that highlight significant astronomical events of 2015–2016 and reflections on science, faith, beauty, and the humanities. Whether it be the flyby of Pluto, the exploration of life on other planets, the discovery of gravitational waves, or the transit of Mercury, this section provides brief reflections on how people of faith can approach these explorations from a perspective that is true to who we are as Catholics and honors the science of these discoveries on its own terms.

This book would not exist without the support of the Vatican Observatory (particularly Br. Guy Consolmagno), the Vatican Observatory Foundation (including my fellow bloggers at *The Catholic Astronomer*), Bishop Callahan of the Diocese of La Crosse, and Peregrino Press. With their assistance and encouragement, I am happy to present to you this reworking of my thoughts from this past year on different areas of faith, astronomy, theology, and science.

A simple question is asked of every book ever written, which is, Why should I read this book? Of the many answers I could give, the one I feel strongest about is also the simplest: God has blessed my life with a fascinating pilgrimage of the mind and heart and strengthened my faith through the exploration of faith and science. I want to share that journey with you! This journey does not promise that all of your questions or doubts will be answered. But it is my hope that it helps you start your own pilgrimage, allowing God to reveal the answers you seek in due time.

Did you ever have a dream that not only came true, but took on a life of its own? I have. And I wish to share that dream with you.

Fr. James Kurzynski

Understanding the nature of Christian
Faith and the nature of Science

The Nature of Christian Faith

A major problem with debates between Christian faith and science is an inability to stay within the parameters of the nature of Christian faith and the nature of science. We read in the Catechism of the Catholic Church that faith is certain since the very Word of God cannot lie and "ten thousand difficulties do not make one doubt."(CCC 157) Read in isolation, one can rush to the presumption that faith is so certain that there is no need to pursue any other intellectual disciplines. However, when we read on in the Catechism, we find that faith seeks understanding, compelling the human person to deepen our understanding of God and the world, through the guidance of the Holy Spirit. This is summarized beautifully by Saint Augustine, "I believe, in order to understand; and I understand, the better to believe." (CCC 158)

These reflections point to Christian faith being a supernatural gift that opens the heart to know, love, and serve God.

As helpful as these reflections may be, we are still seeing faith from within the context of Divine Revelation. This begs the question, how does natural reason (including science) inform faith? When we move onto the next paragraph in the Catechism, our understanding of faith broadens to include natural reason. Let's read the paragraph in its totality.

Faith and science: "Though faith is above reason, there can never be any real discrepancy between faith and reason. Since the same God who reveals mysteries and infuses faith has bestowed the light of reason on the human mind, God cannot deny himself, nor can truth ever contradict truth."

"Consequently, methodical research in all branches of

knowledge, provided it is carried out in a truly scientific manner and does not override moral laws, can never conflict with the faith, because the things of the world and the things of faith derive from the same God. The humble and persevering investigator of the secrets of nature is being led, as it were, by the hand of God in spite of himself, for it is God, the conserver of all things, who made them what they are." (CCC 159)

We can see that the nature of Christian faith is to seek truth, first and foremost, through the Revelation of Jesus Christ and the inspiration of the Holy Spirit. However, Christian faith also calls us to engage the different disciplines of natural reason since the truths found in these disciplines point to the same source as does Divine Revelation. The nature of Christian faith draws upon all disciplines (presuming they are pursuits of real science), affirming a core, fundamental truth that unites all schools of thought, leading to a deeper understanding and appreciation of who God is, how God brought everything into existence, and where God is leading us in our lives. (For an example of how this vision of faith is expressed in education, read Blessed John Henry Newman's text, "The Idea of a University.")

Can such a certain faith also admit times of confusion and contradiction? Absolutely! In fact, faith demands a gap of knowledge in order for the human person to deepen his or her faith, even to the point of times of profound spiritual darkness (for an example of this, read "The Dark Night of Mother Teresa," by Carol Zeleski.)

When it comes to contradictions between faith and science, the response of Christians should not be, "Well, science must be wrong since faith is always right!" Rather, apparent contradictions are to be seen as just that, apparent, and a mutual exploration of truth begins in which faith and reason are in dialogue with one another to deepen our understanding of truth in contrast to trying to "knock each other's block off."

Nature of Science

Let's look at the nature of science. I often find myself frustrated when I hear arguments that can be summarized as, "Faith is about opinions and feelings, while science is about hard facts and objective truth." This drastic oversimplification of both faith and science can breed a false interpretation that if you want to know truth, stick with science; if you want to be sedated by the opiate of the masses, follow religion. As I have grown in my understanding of science as a contributor to *The Catholic Astronomer*, I find, more and more, that to make such claims is not only wrong, but disrespectful of the very nature of both faith and science.

To explore the nature of science, I would invite you to read the National Science Teachers Association's position statement on the nature of science. This summary of the nature of science points to the core parameters in which science is done. Central to this understanding is that scientific knowledge is reliable but tentative, meaning that even the best of research may be modified or discarded in light of new discoveries.

Science employs a number of methods and processes to understand the physical world, but precludes the use of supernatural elements in the production of scientific knowledge. Therefore, science is a limited field that is not designed to explore things beyond the material world.

The summary goes on to affirm that the final goal of science is to understand the natural world for its own sake. So, we can see that even though science is a powerful tool to explore our physical world, there is built into the very nature of science a humble disposition, presuming that the truths arrived at can and will be changed, modified, or completely abandoned. Lastly, since the nature of science is to exclude the supernatural, to make definitive statements about the supernatural such as, "faith is a bunch of opinions," falls outside of the purview of science and is an expression of one's opinion about faith and not something derived from the nature of science itself.

15

A clear articulation of these points is made in the National Academy of Science's open text, *Teaching about Evolution and the Nature of Science*. The text contains two wonderful responses in the "Question and Answer" section in regard to how science views God. When addressing the question, "Can a person believe in God and still accept evolution," the text affirms that many scientists have a deep faith and find no contradiction between their scientific work and their Christian belief.

Another key insight is that many people do not properly understand the difference between scientific knowledge and religious knowledge. Questions of purpose and meaning are not part of scientific investigation but are central questions to people of faith. Therefore, when we try to understand what faith and science have contributed to human history, we need to be aware that there are fundamental differences in their approaches to truth, pointing to two fundamentally different contributions.

The second question asks, "Aren't scientific beliefs based on faith as well?" In response, the text states that the word "faith" means something quite different for faith and science. Christian faith has clear, foundational teachings and tenets that religion is built upon, while science is constantly testing, retesting, modifying, and exploring new data to challenge established principles. The distinction between the unchanging core beliefs of Christians and the constantly changing beliefs of science point to two fundamentally different understandings of the word "belief."

Science has faith in a method.

The final response to the posed question about whether or not scientific beliefs are based on faith ends with a beautiful distinction between the nature of faith and the nature of science.

> Science is a way of knowing about the natural world. It is limited to explaining the natural world through natural causes. Science can say nothing about the supernatural. Whether God exists or not is a question about which science is neutral. (*Teaching about Evolution and the Nature of*

Science. Online Open Book. p. 58)

In light of these reflections, we can see that exploring a debate between Christian faith and science is a difficult feat if the goal is to stay within the parameters of the nature of each. It is not that Christian faith and science are diametrically opposed to one another, but their investigations point to two different ways of knowing and coming to truth. Let's look at how an exchange between Christian faith and science can be done in a healthy manner.

Comparing the Nature of Christian
Faith and the Nature of Science

From our exploration so far, we can derive two important points to help us understand why debates between faith and science often fail:

1.   The nature of Christian faith explores truth that is revealed through the Revelation of Jesus Christ, inspired by the Holy Spirit, and is present through natural reason. There is no need for faith to place limits upon explorations of natural reason (including science) since all truth points to a common origin, namely God.

2.   The nature of Science necessarily places limits upon itself as a tool of understanding the natural world. The question of God's existence is a question that science remains neutral about since scientific method only explores the material world. Therefore, the nature of science does not deal with questions like the divinity of Christ, the existence of the Holy Spirit, and questions of life's meaning and purpose.

From these two affirmations about the nature of Christian faith and science, we can begin to see the difficulties in establishing the groundwork for a healthy debate between faith and science. On the one hand, the nature of Christian faith affirms the necessity of natural reason, not seeing it as a contradiction of or threat to Christian faith. Rather, exploring natural reason is necessary to come to truth (a fact many Christians would do well to revisit).

One may argue that the Church has tried to place limits on scientific study, seeking to scuttle certain advancements in different fields of science (human cloning being a prime example). There is a necessary dialogue that needs to occur about the ethics and morality of the application of science. (This dialogue is not only a concern for Christianity, but for people of all faiths, people of no faith, and the scientific community itself.) However, science as an exploration of truth about the physical world should be allowed the freedom to explore truth, trusting that sincere pursuits of truth will ultimately point to the source of truth.

On the other hand, one may argue that a debate that respects the nature of faith and science naturally places the scientist at a disadvantage, given the limits that science imposes upon itself. What is a scientist to do if a Christian goes on about the truths revealed in Jesus Christ? Does the scientist simply shrug his or her shoulders and say, "Sorry, we don't deal with that."

This challenge, however, can also be a burden on Christians if they truly approach this debate in a spirit of charity. Do we have a debate night where we only talk about natural reason, gutting all vestiges of Christianity from the discussion? Do we have a night where the Christian speaks freely about the different sources of Divine Revelation and the scientist talks about the truths of the natural world, risking an end result of leaving the audience still wondering, "So what is the relationship between faith and science?" I hope you're beginning to see that this question is far more complicated than one may think on the surface. I would take it a step further and pose the question, "Is a debate between Christian faith and science even possible?"

Is a debate on faith and science even possible?

I was walking through a local bookstore the other day and randomly thumbed through texts arguing that a debate between faith and science is not possible. My first inclination was to think that these books might have a good point, given my inner wrestling with this very question. However, the foundational arguments of these books was that Christian faith is so arcane, so immature, so unintelligent, so inflammatory, and so ignorant of the physical sciences that society should lump us all into a ball and kick us out of the discussion!

I was not impressed, especially as these authors went on to espouse themselves as the mature geniuses of our times, almost hinting at a "messianic" mission of which only they possess the golden compass of truth. If you are someone who reads these texts and comes away thinking, "There has to be a better way to explore these questions," you're in good company. These spats, as tempting as they are to engage in, only perpetuate how debates on faith and science should not look. We need a new format if anything healthy can emerge.

A more charitable exploration of this thesis is necessary: Is an actual debate between faith and science possible given the nature of Christian faith and the nature of science? My assertion is that the nature of both faith and science shows us that a debate is not necessary. Mature faith sees natural reason as a necessary means of coming to truth, while honest science affirms that questions about God are not in their spectrum of investigation. The real question should be, "Why do we feel a need to have a debate in the first place?"

A simple answer to this question may be fear. Fear on one side that advancements in science will do harm to Christian faith and fear on the other side that religious belief will create a barrier to advancing our knowledge of science. If a debate between faith and science is to be done, it should be about what has led to the fear and distrust between faith and science, address these concerns, embrace the true nature of Christian faith and science, and then move forward as dialogue partners, not adversaries.

If we are now to shift from debate to dialogue, what should this dialogue look like? To explore this question, let us return to our beginning, revisiting the words of St. John Paul II.

> Faith and reason are like two wings on which the human spirit rises to the contemplation of truth; and God has placed in the human heart a desire to know the truth—in a word, to know himself—so that, by knowing and loving God, men and women may also come to the fullness of truth about themselves. (St. John Paul II, *Fides Et Ratio*.)

The ends of faith and the ends of reason are not always the same, but they can inform one another on their given explorations.

Both faith and reason can teach us that the human person has an inner wiring and longing to know the world that we live in and that this world is multidimensional and multifaceted and that it contains layer upon layer of meaning, purpose, and truth that have captivated the mind for the whole of human history. Even the hardest of atheist scientists can affirm that faith, Judeo-Christian faith in particular, has had a monumental impact on human history, both for good and bad, and even the deepest Christian who is suspicious of the natural sciences can affirm that scientific exploration has greatly impacted modern society, both for good and bad.

Could there be a day when this atheist and this Christian approach one another in a spirit of solidarity, seeing that their common goal of pursuing truth may be a way to charitably explore their differences, or are we doomed to the endless charade of "debates" that have more to do with politics, bitterness, and hatred, laced with sophomoric "gotcha" moments, than a sincere engagement in a dialogue about truth? My prayer is that the words of St. John Paul II will drive a new dialogue of charity between sincere Christians and scientists, bringing the failed "debate" circuit to a necessary end.

Spiritual Exercise: Do you think a debate between faith and reason is possible? Reflect on this question and, together, let us engage in a dialogue between faith and reason, allowing for a moment of ascent toward the source of all truth.

① Does faith make claims that are scientifically falsifiable or verifiable?

② Proportionalist Ethics makes morality depend on outcomes that can be scientifically studied.

EXPLORING OUR ORIGINS

C.S. LEWIS AND THE FIGHT FOR MEANING IN GENESIS

WHAT WOULD IT HAVE been like to be present at the moment of creation? Could our five senses have comprehended the event? Would it have been beautiful, majestic, and grand like gazing upon creation from atop a mountain? Would it have been small, hidden, and quiet like the conception of a child in its mother's womb?

The answers to these questions are elusive because none of us was present at the moment of creation. In fact, it would have been physically impossible for any of us to be there. So how do we explore our genesis?

The next generation of professional telescopes seeks to glimpse the measurable aspects of these events with us safely separated from their ferocity by distance and time. However, this will only explore one aspect of our beginnings, leaving hidden the non-material reality of creation. Still, this exploration will ask us once again, "Where did we come from?"

These thoughts came to me while reading C.S. Lewis' *Reflections on the Psalms*. In one section, Lewis explores how mythology approaches creation in contrast to philosophy and Judeo-Christianity. Lewis asserts that mythology presents creation on a superficial level, similar to enjoying a theatrical play based solely on the sequence of events from the time you enter the theater to when the curtain falls. Although these stories may be enjoyable, we instinctively know that there is something missing when reading creation stories at this level. Any creation story worth reading will do more than make us wait for the story to end. It should leave us wondering, "Where does our story begin?"

Lewis asserts that philosophy and Judeo-Christianity ask fundamentally different questions than does mythology when approaching the subject of creation. Using the metaphor of theater again, philosophy and theology prefer to ask questions about the origin of a thing, wanting to know first how the play originates or comes into being.

This approach implies questions that would pre-date the play's actual performance such as, Who was the author? Who was the director? How was the staging put together? Are there other people behind the curtain we can't see that are essential to the play? Was there a previous text that was edited for this performance? and Do the events carry symbolic significance beyond what the performance of the play indicates?

Lewis explains that when we enjoy a play, most people seldom ask these questions, preferring the simple enjoyment of a performance from beginning to end. However, these questions are important if we want to go beyond a superficial understanding of the play and get to the essence of the message being communicated through the performance. When applying this metaphor to Biblical texts, we can see that asking about the origins of a creation story makes the reader explore the essence of creation that is foundational to the story's existence.

It also points to an important interpretive key when reading

Biblical accounts of creation: Should these stories be read as a sequence of what happened at the moment of creation or should they be read as expressions of what we have come to know about the essence of creation and its Creator, straining to communicate these eternal truths through the limits of human language and experience? (Summary of C.S. Lewis, *Reflecting on the Psalms*. P.79)

The cynic may consider this whole exercise pointless given our inability to obtain firsthand knowledge that "a moment of creation" ever happened, reminding one of the old runaround, "If a tree falls in the woods, but nobody is there to hear it, does it make a sound?"

Similar to how the laws of physics are not contingent upon people being present for them to operate (the tree falling does make a sound), so, too, can we affirm that our very existence points to the fact that creation happened, the metaphorical "tree of our genesis" has fallen and the "sound waves" from that event continue to resonate in creation to this day.

Yet, we still look for the "tree," trying to answer the question that gives the cynic's protest its validity: How did creation happen? Or, perhaps, we should refine our question based on C.S. Lewis' reflections and ask: How does creation happen?

What do we discover when we apply this reflection to the Book of Genesis, specifically the first two creation stories? If we reduce our interpretation of Genesis to the events of the story, we can quickly get hung up on the inconsistencies between the text's presentation of creation and what we know scientifically of the created world. However, if we ask how the story of Genesis came to be, we would then need to compare it to other creation stories of its time.

Sometimes, people get nervous with this approach, fearing that it will reduce the Bible to merely one text among many that contain "opinions" about God (unfortunately, some have tried to do this very thing). However, authentic historical exploration of the text is meant to ask questions about its genre, the audience who will hear it, other ancient texts that may have contributed or are in opposition to the Bible, the cultural context of the people of the time, and many other aspects ranging from the original languages that were spoken to the nature of the people's worship.

One of the common historical comparisons is to look at Genesis 1–3:24 in contrast to a series of ancient scrolls called the *Enuma Elish.* When comparing these texts, we find clear differences that point to two fundamentally different concepts of creation.

One of the clearest distinctions is that Genesis points to a disposition of peace between God and creation in that God creates through a Word Act in contrast to the *Enuma Elish* that presents a polytheism of jealous gods being at war with one another, bringing about creation through violence (this being a trend common to many mythologies). This distinction brings us to a fundamental truth of how creation happens according to Genesis: Creation is accomplished through love from a God who is not at war with us or Himself. (Summary of *The Collegeville Bible Commentary: Old Testament.* P.38)

Continuing this distinction, if we read Genesis 1:14–16 on a superficial level, we can get stuck on the visual of the sun and the moon traveling around the earth, contrary to our current understanding of the Solar System. If we ask, instead, how this "day of creation" originated and what it is trying to communicate, we find that the sun and moon are called "lights" to clarify that Judeo-Christian belief sees them as part of the material world in contrast to the pagan religions of the time that worshiped both sun and moon as gods.

Much more can be explored in this "comparative religions" approach to the texts (such as the Christian affirmation of free will and the pagan disposition toward predestination), but, for the sake of time, we will simply point out that asking how the stories of Genesis originate can reveal to us truths of who God is, how God relates to us, and how we are to relate to God and the world. (Summary of footnotes from Genesis 1:16, *The Jerusalem Bible.*)

When we explore the origins of these stories, it can feel like we're part of an exciting mystery, pealing layer after layer of the "onion" of truth. When I bought my copy of *The Catholic Study Bible* back in college, I, unfortunately, skipped the background section, eagerly wanting to get into the text. Thinking the introductions were boring, I did myself a true disservice by not understanding the context of the stories I was reading. Instead, I should have struggled to learn the different sources of Genesis we call the Yahwist (J), the Elohist (E), the Deuteronomist (D), and the Priestly Source (P).

I would have learned that the first creation narrative in Genesis 1–2:3 (a "P" source) is actually "younger" than the second creation narrative in Genesis 2:4–3:24 (a "J" source). Therefore, Genesis 1–2:3 is composed in light of Genesis 2:4–3:24, emphasizing in the "first creation story" what can get lost in the "second creation story:" That the human person is created in God's image and likeness, we are fundamentally good, despite our sinfulness, and live in a fundamentally good creation which, despite us being bound to labor and toil upon after the fall, is to be approached from a standpoint of stewardship and not exploitation. Further, the poetic symmetry in Genesis 1–2:3 in which day one corresponds to day four, day two corresponds with day five, day three corresponds to day six (look it up to see what I mean), displays a literary structure that is more poetic and hymn-like, pointing to the "end" of creation as the "eternal day" of the Sabbath. (Summary of The Catholic Study Bible: New American Bible. RG47–RG61)

This final reflection on how Genesis points creation to the Sabbath can tease out another layer of exploration: Is the creation story about history or is it a text about liturgy? This question of creation and worship was a central theme to a collection of Pope Emeritus Benedict XVI's homilies (back when he was Joseph Cardinal Ratzinger) that appears in the work, "In the Beginning…' A Catholic Understanding of the Story of Creation and the Fall."

Gen 1 : History or Liturgy?

The creation accounts of all civilizations point to the fact that the universe exists for worship and for the glorification of God. This cultural unity with respect to the deepest human questions is something very precious. In my conversations with African and Asian bishops, particularly at episcopal synods,

it becomes clear to me time and time again, often in striking ways, how there is in the great traditions of the peoples a
oneness on the deepest level with biblical faith. In these traditions there is preserved a primordial knowledge, which serves
as a guidepost and which links the great cultures, and that
an increasing scientific know-how is preventing us from being
aware of the fact of creation. (Pope Emeritus Benedict XVI,
"In the beginning... p.28)

Having gone through this brief (and insufficient) explanation of
Genesis, ending with Pope Emeritus Benedict's thoughts on a creep-
ing "scientific know-how" that is preventing us from being aware of
the fact of creation, let us return to C.S. Lewis' play. Can we not see
that trying to impose a superficial understanding of creation upon
the book Genesis is missing the central point the text is trying to
communicate? Can we not see that to understand these texts properly, they need to be read in light of how creation originates instead
of fixating on when the "drama" begins and ends on the stage of our
imagination? Do we not hear a call to all of creation to worship, in
love, the God who loved us into existence? And do we not see, in
this Inspired text from Genesis, the very creative and fecund nature
of God, allowing us the sacred privilege to contemplate these divine
mysteries and participate in God's creation?

Is Genesis in opposition to science? Read superficially, yes, but
not if you explore how this text came to be and the theology behind
why it is considered an Inspired text. Put another way, Genesis is a
work of theology, not a work of science.

Will we ever find a way to experience the moment of creation?
Unless it be given by God's grace, my guess is no. However, let us
walk together in charity, trying to understand why our world was
created so that we may someday experience, God willing, that the
end our beginnings points us toward the Kingdom of God.

Spiritual Exercise: What do you emphasize when reading creation stories—The unfolding of events, the meaning behind the events, or both?

STEPHEN HAWKING

DIDN'T DR. HAWKING DISPROVE THE NEED FOR GOD?

A GREAT DEAL OF SCUTTLE has been raised these past few years among faith, philosophy, and science over a simple question: Can something come from nothing?

Stephen Hawking, in his book (coauthored by Leonard Mlodinow) titled *The Grand Design*, argues that it can, leaving no need for a God to bring things into existence. Many in theology, philosophy, and science have said that Hawking's claims are not accurate and there is a "rewrite" going on in regard to what it means to create "from nothing" (Ex Nihilo). In this section, I will briefly explore the meaning of the word "nothing" as it is used in philosophy, theology, and in Hawking's presentation of creation in the book *The Grand Design* to help us understand the difference between the Christian understanding of *Ex Nihilo* and what Hawking presents as creation Ex Nihilo.

To begin, let's understand the philosophical and theological concept of creation from "nothing." I believe many Christians don't properly understand what this concept means. Yes, we may have heard numerous times since our youth that God created all things from nothing, but have we taken the time to ask the difficult questions to understand what that actually means?

First, does it mean that you begin with a state of being we call "nothing" that is then changed into "something?" Philosophically and theologically, this simple idea is in error and needs correction. To claim that there is a "state of being we call nothing" changes nothing into something. Therefore, nothing is not a state of being, it isn't a pre-existing material we don't understand, and it isn't a measurable reality outside of space and time.

All of these designations imply the existence of something. Therefore, creation from nothing is not a "change" from one state to another, but the reality that creation happened and is ongoing. The best explanation of nothing is, well, nothing! Any attempt to identify nothing as a reality violates its very definition and turns "nothing" into "something." A helpful explanation of this through the eyes of St. Thomas Aquinas can be found in the writings of William Carroll from the University of Oxford.

> Creation, on the other hand, is the radical causing of the whole existence of whatever exists. To cause completely something to exist is not to produce a change in something, is not to work on or with some existing material. If, in producing something new, an agent were to use something already existing, the agent would not be the complete cause of the new thing. But such complete causing is precisely what creation is. To build a house or paint a picture involves working with existing materials and either action is radically different from creation. To create is to cause existence, and all things are totally dependent upon a Creator for the very fact that they are.

Dr. Carroll goes on:

> Aquinas shows that there are two related senses of creation, one philosophical, the other theological. The philosophical sense discloses the metaphysical dependence of everything on God as cause. The theological sense of creation, although much richer, nevertheless incorporates all that philosophy teaches and adds as well that the universe is temporally finite. (William Carroll, "Creation, Evolution, and Thomas Aquinas." Online text)

To apply this reflection by Carroll, let's explore a question I receive, at times, as a priest: "Father, was the Big Bang the beginning of creation?" The honest answer to the question is "I don't know," but logic seems to tell me that the deeper answer lies somewhere between "probably not" and "no." Why? Let's explore this by asking another question: Even though we affirm that the Big Bang is not incompatible with the Bible, must we have a Big Bang to affirm God as Creator? No, we don't.

There are numerous ways we could comprehend the coming into existence of things that would differ from the Big Bang. Let's ask another question: If there was something before the Big Bang (literally meaning anything), would this contradict the idea of God as Creator? Again, no, it would not contradict the idea of God as Creator because we are not bound by faith to say that the Big Bang was the beginning of creation and, in addition to the physical world, we believe in a non-physical, yet created existence that would include the angelic realm.

There are a number of scenarios one can imagine that would include something in creation before the Big Bang that would not contradict faith. Lastly, if the natural sciences failed to discover a "moment of creation," would this contradict the idea of God as Creator? No, it would not contradict the idea of God as Creator since, as we stated before, creation is not a change from "nothing" into

Is there such a broad divide b/w the two?

"something," but rather creation is the fact that things have come into existence and is a question of metaphysics and not of science. To point this out is in no way disrespectful of science since science places limitations upon itself and remains neutral on the question of God and anything non-material.

This is a rather brief exploration into the question of the philosophical and theological understanding of creation Ex Nihilo. To delve into the topic more deeply, read Question 46 from Thomas Aquinas' Summa Theologica, Thomas Aquinas' Summa Contra Gentiles Book 2, Paragraphs 15–21, or Aquinas on Creation and the Metaphysical Foundations of Science by William Carroll. The foundational points we can establish thus far are the following:

1. "Nothing" is not a type alternate reality. If it were, it would cease to be nothing and become something.

2. To speak of creation Ex Nihilo does not mean that we have a "change" in substance from nothing to something. Creation is not an act of change, but rather the fact that creation happened and is ongoing.

3. Creation Ex Nihilo, as defined by philosophy and theology, does not present itself as a question of science, but is a question of metaphysics, a field of which science itself does not address.

With this as our starting point, we will now look at Hawking's book, *The Grand Design*. I want to begin by being upfront that I have been a fan of Hawking's writing ever since college. His book, *A Brief History of Time*, was one of the few books I read multiple times in my youth. One of the reasons the book spoke so deeply to me was the author's apparent openness to the idea of God in relation to creation. The quote from Hawking that so gripped me and many others was the last sentence of his conclusion.

… (I)f we do discover a complete theory, it should in time be understandable in broad principle by everyone, not just a few scientists. Then we shall all, philosophers, scientists, and just ordinary people, be able to take part in the discussion of the question of why it is that we and the universe exist. If we find the answer to that, it would be the ultimate triumph of human reason – for then we would know the mind of God. (Stephen Hawking, *A Brief History of Time*. p. 193)

This sentence alone made me want to reread his text so I could come to this conclusion a second and third time. It gave birth to the hope that a Theory of Everything (ToE) could also include God and we would finally have a universally accessible and acceptable notion of the truth about the world and about God. Later on in life, however, I began to learn that this very thought was flawed due to its reduction of everything to a material definition. In light of this, I shouldn't have been surprised that Hawking has now arrived at a thesis that excludes God from the elusive "ToE."

Because gravity shapes space and time, it allows space-time to be locally stable but globally unstable. On the scale of the entire universe, the positive energy of the matter can be balanced by the negative gravitational energy, and so there is no restriction on the creation of whole universes. Because there is a law like gravity, the universe can and will create itself from nothing… Spontaneous creation is the reason there is something rather than nothing, why the universe exists, why we exist. It is not necessary to invoke God to light the blue touch paper (a fuse to something that explodes) and set the universe going. (Stephen Hawking, Leonard Mlodinow, *The Grand Design*. P. 180)

Now, for good or for ill, most reflections reduce Hawking and Mlodinow's text to this one quote. The entire text itself is brief, but makes some rather audacious claims such as Dr. Hawking's assertion that "philosophy is dead." The irony of this claim by Hawking is twofold: 1) As we saw in the block quote from his work, *A Brief History of Time*, one of Hawking's goals was to make his theory accessible to philosophers and ordinary people – not replace philosophy; and, 2) Hawking himself provides a philosophical framework that, if analyzed by philosophy, shows itself to repeat basic intellectual errors of the past (material reductionism as the main error). In order to explore this, we need to understand the core scientific ideas presented in Hawking's text.

I am not a scientist. Therefore, I will call on the help of Dr. Stephen Barr, Physics and Astronomy professor at the University of Delaware and a routine contributor to the Catholic Periodical *First Things*, to explain what Hawking is presenting as a vision of the universe and how things come into existence.

Dr. Barr nicely lays out the thought of Hawking in a piece titled, "Much ado about 'nothing': Stephen Hawking and the Self-Creating Universe." Barr begins by explaining that Hawking's notion of "creation from nothing" isn't new to science and can be traced back to a theory of things coming into existence through "quantum fluctuations," dating back to 1982 in the work of Alexander Vilenkin. Dr. Barr writes that there are great difficulties that need to be overcome to actually verify that these theories exist in reality and perhaps would be better explained as scenarios of creation.

In this view of the universe, Barr explains that, from the starting point of one particular structure, we can think of quantum fluctuations as universes going in and out of existence similar to how a balloon can be pinched and twisted to create "smaller balloons" from one larger balloon. The point being is that we can conceptualize a state of a universe in which there is "nothing" or a "no universe state" in one of the smaller "balloons" that then grows into an actual universe. Therefore, you would have one, overarching "system of universes" with a number of "smaller universes" coming in and out of existence.

Some states of the system of universes would correspond to just one universe being in existence; others to two universes, and so on. And there would also be a state with no universe in existence. The dramatic possibility Hawking is considering (and many others before him) is that such a system might make a transition from its "no-universe state" to a state with one or more universes. (Stephen M. Barr, "Much ado about 'nothing:' Stephen Hawking and the Self-Creating Universe," *First Things*. [9/10/2010] Online)

I want to emphasize that there is nothing in this explanation of the universe that I feel a need to critique from the standpoint of whether or not the science itself is accurate. In fact, I think it would be fascinating and wonderful if this turns out to be how our universe actually works mechanically. The question I wish to ask is this: "Does the idea of going from a 'no-universe state' to a 'state with one or more universes' constitute an understanding of creation Ex Nihilo that not only speaks to, but also replaces the classic philosophical and theological definition of creation from nothing?"

The answer should be clear that, no, this does not represent, replace, or disprove the philosophical and theological understanding of creation Ex Nihilo. Rather, it reflects the passing of one reality, the "no-universe state," to another reality, "a state with one or more universes." This, alone, violates the philosophical principle that creation Ex Nihilo is not a change, but the "radical causing of the whole existence of whatever exists" to borrow Carroll's language from earlier.

What we learn from exploring Hawking's argument for creation Ex Nihilo is that it is far different from what philosophy and theology understand as creation from nothing. Further, when exploring the thought of Hawking and of the philosophical/theological understanding of creation, what we find in the book *The Grand Design* is not a watershed moment of being able to replace one idea of creation from nothing with another. Rather, a study of these ideas reveals that Hawking's text reflects a classic misunderstanding of what creation

Ex Nihilo means. If this distinction were to be explored in his text, Hawking's understanding of creation Ex Nihilo would reveal itself as not reflecting creation from nothing, but rather the coming into existence of things from one state of being to another state of being.

I wish to emphasize that, as I stated earlier, this critique in no way requires us to question the science that Hawking presents. It would not surprise me in the least if some day, even with the difficulties mentioned by Barr in proving this scenario, that Hawking's science may be proven correct. And if it is, it will not present itself as a disproof of God's existence or a sufficient replacement for the classical understanding of creation Ex Nihilo.

FOUNDATIONS OF CREATION

WHAT IS CREATION EX NIHILO AND CREATIO CONTINUA?

IN LAYING OUR FOUNDATIONS, let's continue to look at Creatio Ex Nihilo (Creation from Nothing) by contrasting it with another important aspect of creation called Creatio Continua (Continual or Ongoing Creation). These two understandings of the relationship between God and creation go back to the earliest writings of Christianity after the life, death, and resurrection of Jesus Christ. These theologies are developments upon the core understanding of creation found in Genesis. Here is a summary of those points from *The New Dictionary of Theology.*

1. The whole of creation was brought into existence by a free, loving act of God.

2. When we explore the genre of Genesis, it is not a book of history or science, but is of the same genre of the ancient creation stories of its time.

3. Creation is fundamentally good and expresses the continual goodness of God to creation.

4. God is not the source of evil, but evil is the absence (or privation) of the good in the world.

5. Creation is made for the human person and humanity is called in return to be good stewards of creation. (Take from "Creation," *The New Dictionary of Theology*. p. 247–248)

From these points of interpretation comes our understanding of the fundamental nature of creation and creation's dependence upon God. However, a clear Biblical reference to a theology of creatio ex nihilo does not appear until the second book of Maccabees.

I beg you, child, to look at the heavens and the earth and see all that is in them; then you will know that God did not make them out of existing things. In the same way humankind came into existence. (2 Maccabees 7:28)

The understanding of creatio ex nihilo is well known and embraced by many Christians (which we explored in our previous section). However, what is less understood is the next logical question that follows from creatio ex nihilo: If God created all things from nothing, did God create everything in one instant or is God's creative act ongoing?

This question opens the door to many other foundational questions about God and creation: How do we account for the coming into existence of new species throughout history? Why would God allow certain species to become extinct? How do we understand change in relationship to time? To help us answer these questions, we will enlist the help of the Eastern Church Fathers and St. Augustine.

In regard to the Eastern Church Fathers, two significant authors we can draw upon are Clement of Alexandria (150A.D. – 215A.D.) and Origen (184A.D. – 253A.D.). In addition to embracing creatio ex nihilo, Clement also introduced an understanding of a "continual act of creation" called the creatio continua. This understanding was that God's act of creation did not cease at the first moments of existence, but rather the act of creation is ongoing with things constantly coming into existence.

Origen takes this understanding of creatio continua and places it within a Trinitarian framework, developing his theology of "exitus-reditus" in which all of creation comes from God (the exitus) and ultimately returns to God (the reditus). Therefore, our understanding of creatio ex nihilo and creatio continua includes an exploration of why things come into existence in addition to exploring philosophically how things come into existence. If all things come from God and return to God, then there is a reason this "going out" and "coming in" relationship exists. Also, this continual act of creation helps us understand that it is necessary for certain things to exist at certain times of history. (Example: There is a reason I exist at this point of history and did not exist at the time of Jesus Christ.)

The idea of a continual creation implies a discussion about the role of time in creation. To help us understand this relationship, we look to the Church Father Augustine. Augustine affirms that all things receive their being from God, but he also adds a fascinating reflection on the relationship between time and the universe. Augustine argues that time does not have a spatial relationship with creation, but rather is a function and measurement of change.

I was reminded of this while being interviewed by Bob Berman from Slooh, a community of online observatories. While discussing the scientific and theological understandings of time, Dr. Berman shared with me that the prevailing theory of modern physics is that time is an illusion and that the universe is eternal in nature. When talking about this interview with one of my student parishioners from the University of Wisconsin-Stout, he added that time is merely the study of decay, creating the illusion of time. I don't understand physics well enough to definitely state that this theory is in concert with the thought of Augustine, but in both instances I find it interesting that Augustine and modern physics see time as a function of how the world changes.

A key difference between modern physics and Augustine is that Augustine's understanding of change also implies the change of our spiritual lives and our relationship with God. What we find in the early Church is a tantalizing exploration into how and why things come into existence (from a philosophical and theological standpoint). This change does not happen in a moment, but is an ongoing process of continual creation by God. Therefore, we once again find in the early Church a clear framework to argue that the unfolding of God's plan of salvation (the Economy of God) implies a necessary change in the world.

Spiritual Exercise: How do you view the relationship between God, creation, time, and your daily life? Is your view of God as a distant reality that brought all things into existence, but now is absent from your daily life? Do you see God's creative act as dynamic and ongoing, allowing new realities and possibilities to be brought into our world? Pray with these questions and, as we experience the change that comes with the passing of time, may we open our hearts to allow God's dynamic creative act to renew us, helping us to become an image of God's love in the world.

*Note: The majority of this reflection is a summary from *The New Dictionary of Theology*. Komonchak (editor), (Liturgical Press 1987). 247–250

THE CATHOLIC CHURCH AND EVOLUTION

ISN'T THE CHURCH AGAINST EVOLUTION?

With all the discussion in previous sections about creation, continual creation, and change in creation, one may wonder how the change we are speaking of in theology relates to the change we speak of in the theory of evolution. This next section will explore this relationship.

To begin, let's look at the language of evolution by asking the question, "Can a faithful Christian believe in evolution?" Despite a number of Papal statements in support of evolution, there are some who still question whether or not Christianity and evolution are compatible. An example that many will cite is Christoph Cardinal Schönborn's editorial in the *July 7, 2005,* edition of *The New York Times.* In this piece, Cardinal Schönborn states the following:

Evolution in the sense of common ancestry might be true,

but evolution in the neo-Darwinian sense -- an unguided, unplanned process of random variation and natural selection -- is not. Any system of thought that denies or seeks to explain away the overwhelming evidence for design in biology is ideology, not science. (Christoph Schonborn, "Finding Design in Nature," The New York Times. July 7, 2005. Online)

In this piece, prior to the publication of his book on faith and science titled, *Chance or Purpose? Creation, Evolution and a Rational Faith*, it seems very clear that Cardinal Schönborn is arguing that Christianity and evolution are not compatible. As a young, non-scientist priest, I, too, presumed that, given the well-earned reputation of Cardinal Schönborn as one of the brightest intellects among the College of Cardinals, that perhaps I misunderstood St. John Paul II's message to the Pontifical Academy of Sciences that displayed a great openness to evolution. Was what I presumed to be compatible now incompatible?

Once again, let's draw from the wisdom of Dr. Barr to clarify this point of tension. In his piece titled, The Design of Evolution, Dr. Barr critiques Cardinal Schönborn's statement in the *New York Times*, arguing that the Cardinal slipped into a confusion of language in regard to the use of the word "random." Dr. Barr explains that Cardinal Schönborn clearly connects the idea of random variation and natural selection with the presumption that these are "unguided, unplanned" processes, divorced from divine providence. Dr. Barr argues that, from the standpoint of philosophy and theology, if the meaning of the word "random" meant unguided, unplanned, and meaningless, then Cardinal Schönborn's critique would be correct.

Dr. Barr argues that the scientific understanding of the word "random" does not point to something that lacks meaning, but rather points to a gap in our knowledge that needs to be explored. If a biologist, for example, were to use the word "random" to argue that the evolution of things in this world was meaningless, this scientist would be using the word in a rather careless fashion or would possibly be promoting a personal philosophical worldview. Instead, the word "random" is better understood as "uncorrelated." Dr. Barr provides a simple analogy to demonstrate the difference between "uncorrelated" and "unplanned/meaningless."

> My children like to observe the license plates of the cars that pass us on the highway, to see which states they are from. The sequence of states exhibits a degree of randomness: a car from Kentucky, then New Jersey, then Florida, and so on because the cars are uncorrelated: Knowing where one car comes from tells us nothing about where the next one comes from. And yet, each car comes to that place at that time for a reason. Each trip is planned, each guided by some map and schedule. Each driver's trip fits into the story of his life in some intelligible way, though the story of these drivers' lives are not usually closely correlated with the other drivers' lives. (Stephen M. Barr, "The Design of Evolution," First Things. October 2005. Online)

Dr. Barr provides other analogies and reflections, each pointing to the same end: Random, in the scientific sense, does not disprove divine providence. Further, the theological vision of divine providence does not need to exclude evolutionary processes that are random because our understanding of the world is deeply rooted in contingency or, to quote Dr. Barr, what normal people call "chance."

> Communion and Stewardship settles this point. "Many neo-Darwinian scientists, as well as some of their critics, have concluded that if evolution is a radically contingent materialistic process driven by natural selection and random genetic variation, then there can be no place in it for divine providential causality... But it is important to note that, according to the Catholic understanding of divine causality, true contingency in the created order is not incompatible with a purposeful di-

vine providence. Divine causality and created causality radically differ in kind and not only in degree. Thus, even the outcome of a purely contingent natural process can nonetheless fall within God's providential plan... In the Catholic perspective, neo-Darwinians who adduce random genetic variation and natural selection as evidence that the process of evolution is absolutely unguided are straying beyond what can be demonstrated by science." (Stephen M. Barr, "The Design of Evolution.")

My takeaway from this is that just because something is random it does not mean that it is meaningless. We may currently lack the understanding of why certain natural phenomenon occur, but science itself presumes that, with time and study, those "gaps" of knowledge will be closed. In fact, for science to say that "random variation" is meaningless seems to present similar problems as arguing for God's existence by looking for "irreducibly complexity" in nature.

If a scientist were to observe a random process of nature and declare it meaningless, but later this process emerged as an essential key in understanding the development of other species, wouldn't this scientist experience professional embarrassment similar to a person of faith who would declare that God could be proven through something in nature that was irreducibly complex, unexplained by evolution, only to have time and research reveal that their example was not so complex after all and easily explained by evolution? I am not a good enough student of either school of thought to offer a definitive answer to my own question. Nevertheless, as I try to understand the logic of each, there seems to be a similarity in the issues that can arise for both ideologies.

These questions aside, a key point I learned from this exploration was that even the brightest of intellects who have the best of intentions to communicate the truth can stumble when there is confusion of how language functions between different intellectual disciplines, which underscores the need for a clear communication of meaning in the words we use between faith and science.

If these types of misunderstandings can occur between professional scientists and Cardinal theologians, how much more for those of us who are not scientists or theologians, seeking to discover a language that gives voice to truth, trying to understand the material world and non-material existence?

Can the perceived tension between evolution and God as Creator be fully addressed by clarifying a breakdown in communication over the word "random?" Of course, the answer is no. As is the case with many strained relationships, there isn't just one issue that leads to a breakup. Nevertheless, I do feel that these explorations are needed for both believer and non-believer if we are to move away from faith and science as adversaries and toward a position of faith and science as dialogue partners in search of truth.

As mentioned earlier, Dr. Barr refers to a document from the International Theological Commission (headed at the time by then Cardinal Ratzinger, now Pope Emeritus Benedict XVI) titled, "Communion and Stewardship: Human Persons Created in the Image of God." In this document, there is a careful analysis of the relationship between faith and science. I will conclude by providing a paragraph from this document for your consideration.

With respect to the evolution of conditions favorable to the emergence of life, Catholic tradition affirms that, as universal transcendent cause, God is the cause not only of existence, but also the cause of causes. God's action does not displace or supplant the activity of creaturely causes, but enables them to act according to their natures and, nonetheless, to bring about the ends he intends. In freely willing to create and conserve the universe, God wills to activate and to sustain in act all those secondary causes whose activity contributes to the unfolding of the natural order which he intends to produce. Through the activity of natural causes, God causes to arise those conditions required for the emergence and support of living organisms, and, furthermore, for their reproduction and differentiation. Although there is scientific debate about the degree of purposiveness or design operative and empirically observable in these developments, they have de facto favored the emergence and flourishing of life. Catholic theologians can see in such reasoning support for the affirmation entailed by faith in divine creation and divine providence. In the providential design of creation, the triune God intended not only to make a place for human beings in the universe but also, and ultimately, to make room for them in his own trinitarian life. ("Communion and Stewardship" *International Theological Commission.* 68)

EARLY LANGUAGE OF SPIRITUAL "EVOLVING"

IRENAEUS OF LYONS AND THE OPULENCE OF GOD

I N SECTION ONE, WE have explored a foundation to move away from a "science vs. religion" attitude try to discover a "science *and* religion" stance in which theology and science can be seen as two great disciplines exploring truth in two very different ways. We have explored the ideas of creation from nothing (ex nihilo), continual or ongoing creation (creatio continua), and evolution. To conclude this first section, we now turn to one of the earliest examples of a theological vision that intuits the movements of God's grace with the change that we view in the natural world. The vision we will explore is that of St. Irenaeus of Lyons and his understanding of the Opulence of God.

Irenaeus of Lyons is one of the most significant early Church Fathers. Scholarly opinion places his birth at about 120 AD (give or take five years). He was one of the strongest theological voices to address the problem of Gnosticism in the early Church. (I will touch on Gnosticism more in the sections ahead.) Another part of his significance is that he was a disciple of St. Polycarp, who was a disciple of John the Apostle. Therefore, Irenaeus provides us with an important window into the apostolic age.

In regard to God and creation, Irenaeus used a number of striking metaphors to explore this relationship such as God as Master Architect and God as Master Artist. These and other images of God were central to his theological understanding of the Economy of God, or the slow working out of the plan of salvation. A core aspect of the Economy of God is the development of the soul, moving from infancy to maturity.

> Now, having made man lord of the earth and all things in it, He (God) secretly appointed him lord also of those who were servants in it. They however were in their perfection; but the lord, that is, man, was (but) small; for he was a child; and it was necessary that he should grow, and so come to (his) perfection. And, that he might have his nourishment and growth with festive and dainty meats, He prepared him a place better than this world, excelling in air, beauty, light, food, plants, fruit, water, and all other necessaries of life, and its name is Paradise. And so fair and good was this Paradise, that the Word of God continually resorted thither, and walked and talked with the man, figuring beforehand the things that should be in the future, (namely) that He should dwell with him and talk with him, and should be with men, teaching them righteousness. But man was a child, not yet having his understanding perfected; wherefore also he was easily led astray by the deceiver. (Irenaeus of Lyons, *The Demonstration of the Apostolic Preaching.* 12)

This language is obviously tied with Irenaeus' understanding of sin, moving the soul from immaturity to maturity. This early language lacks the doctrinal precision of St. Augustine's and St. Thomas's writings on Original Sin, but for the sake of this reflection, it is sufficient to say that Irenaeus' understanding of the development of the human person displays an obvious dependency upon God to bring the soul to full maturity so as to avoid "the deceiver." At this point, we can pick up the theme implied in the title of this section: The Opulence of God.

When one hears the word "opulence," the mind initially races toward riches. In the secular world, opulence can take on a very negative sense, associating the word with excessive wealth, exploitation of the poor, and economic systems that fail to care for the most vulnerable of society. In the Biblical sense, the opulence of God means something very different. One of the central Biblical ideas of the opulence of God is found in Second Corinthians when St. Paul reflects upon the self-emptying of God's love in Jesus Christ.

> For you know the gracious act of our Lord Jesus Christ, that for your sake he became poor although he was rich, so that by his poverty you might become rich. (2 Corinthians 8:9)

This image of the "riches of God" being emptied in Christ, voluntarily taking on poverty so that we may be made "rich," is rightly understood as God sparing nothing for humanity. In Eric Osborn's text simply titled, Irenaeus of Lyons, he provides a concise explanation of how Irenaeus explores the Biblical framework of the opulence of God. Osborn argues that from God's power, goodness, and wisdom comes forth all that exists, while God's love and kindness sustains this creation. Therefore, creation itself is an expression of the opulence of God, shown through beauty and the maturation (or evolution) of all things. Osborn references book four of St. Irenaeus' work, Against Heresies, to demonstrate this point.

> With God there are simultaneously exhibited power, wisdom, and goodness. His power and goodness [appear] in

this, that of His own will He called into being and fashioned things having no previous existence; His wisdom [is shown] in His having made created things parts of one harmonious and consistent whole; and those things which, through His super-eminent kindness, receive growth and a long period of existence, do reflect the glory of the uncreated One, of that God who bestows what is good ungrudgingly. For from the very fact of these things having been created, [it follows] that they are not uncreated; but by their continuing in being throughout a long course of ages, they shall receive a faculty of the Uncreated, through the gratuitous bestowal of eternal existence upon them by God. And thus in all things God has the pre-eminence, who alone is uncreated, the first of all things, and the primary cause of the existence of all, while all other things remain under God's subjection. But being in subjection to God is continuance in immortality, and immortality is the glory of the uncreated One. By this arrangement, therefore, and these harmonies, and a sequence of this nature, man, a created and organized being, is rendered after the image and likeness of the uncreated God, - the Father planning everything well and giving His commands, the Son carrying these into execution and performing the work of creating, and the Spirit nourishing and increasing [what is made], but man making progress day by day, and ascending towards the perfect, that is, approximating to the uncreated One. For the Uncreated is perfect, that is, God. Now it was necessary that man should in the first instance be created; and having been created, should receive growth; and having received growth, should be strengthened; and having been strengthened, should abound; and having abounded, should recover [from the disease of sin]; and having recovered, should be glorified; and being glorified, should see his Lord. For God is He who is yet to be seen, and the beholding of God is productive of immortality, but immortality renders one nigh unto God. (Irenaeus of Lyons, *Against Heresies.* Book 4.38.3)

[handwritten margin note: Heraclitus: Everything is in flux "No stepping in the same river twice"]

[handwritten margin note: Parmenides: Anything that is has always been]

This theological vision demonstrates a harmonious union between God and creation, arguing that God has not only brought all things into existence, but also provides grace to sustain creation and allows it to grow. I find it fascinating that one of the earliest, non-scientific, post-resurrection theologies of the relationship between God and creation demonstrates a deep openness to a vision of the world that not only can evolve, but must evolve as part of the Economy of God.

Of course, we need to be careful not to jump to the conclusion that Irenaeus' vision is completely consistent with modern science, but it does display a basic theological and philosophical intuition that affirms the two-fold reality of an unchanging God whose creation is constantly changing. This dynamic, which finds its historical roots in the writings of Heraclitus and Parmenides, inspired later theologians to further develop this relationship between God and creation, finding its high point in Blessed John Henry Newman when he penned, To live is to change, and to be perfect is to have changed often!

This vision of the opulence of God also contains an implicit ethic. If the "riches of God" implies that all has been given by God, sparing nothing out of love for humanity, then we have a responsibility to embrace a similar "self-emptying" ethic to ensure and affirm the dignity of all of creation.

If the understanding of the "riches of God" was nothing more than a "super-being" hoarding the gold of the world, begrudgingly giving small portions of these riches to humanity from time to time, then we would find justification for the same type of exploitation and greed toward our neighbor. However, the Christian understanding of God is of one whose "treasure-trove" is completely empty, giving all in a free act of love.

From this perspective, we see that every part of creation is gift, should be treated as such, and we are called to approach creation with a similar self-emptying love to build up the common good. Because creation changes and develops, we need to understand how it changes and develops. This ethic of free, self-giving love calls us to build up one another in Christ and care for our common home to protect and promote human dignity. Put another way, the opulence of God is meant to realize one of the most oft-quoted ideas of St. Irenaeus: The glory of God is man fully alive, and the life of man is the vision of God! (Saint Irenaeus, *Against Heresies*, Book 4.20.7)

Spiritual Exercise: As we come to the conclusion of Section One, how are we being calling to be good stewards of "God's riches?" Does our attitude reflect a gratuitous God that spares nothing out of love for us or a begrudging king who is prone to hoarding and exploitation? Pray with this question of the opulence of God and, together, may we embrace a heart willing to imitate the generous love of Jesus Christ.

CONCLUSION

SECTION ONE

WE HAVE EXPLORED THE limitations that science places upon itself, remaining neutral on questions of God and metaphysics, in contrast to theology that affirms both Divine Revelation and reason point to a mutual exploration of truth that should be expressed more as a dialogue than a debate.

Foundational to this dialogue is to understand that faith asks fundamentally different questions than science does in their common exploration of truth. Nevertheless, these different approaches still seek out truth about the world we live in, pointing to an ultimate source of creation.

Some, such as Steve Hawking, have claimed, that this exploration leads to an understanding of creation that is in no need of a Creator, but can be explained through natural forces, such as the law of gravity. However, we came to see that what Dr. Hawking presents as creation from nothing does not demonstrate a watershed moment of showing philosophy and theology to be dead, but rather repeats errors of logic that are as old as Plato and Aristotle. Once addressing these errors, we can see that the true sense of creation from nothing points to a dynamic God whose creative act is ongoing.

This ongoing creation by God is not inconsistent with evolution when the scientific definition of evolution is understood in its proper sense. In fact, as is demonstrated in Irenaeus of Lyons, a changing, evolving world can be understood as part of the Economy of God, or God's ongoing work of salvation for all the world, seeing all of creation going through a type of spiritual maturation. From these foundations, let us move on to explore big thinkers of faith and science and how their thoughts have contributed to the ongoing dialogue between faith and science.

INTRODUCTION

SECTION TWO

Having laid a basic foundation for understanding the proper relationship between faith and science, we will now explore several big thinkers, some from faith and some from science, to better understand a few of the biggest scientific ideas of our time and how faith can approach these ideas.

To begin, we will reflect on the work of "priest scientists" by the names of Monsignor Georges Lemaitre, "father" of the Big Bang, and Father Stanley Jaki, the priest who questioned the plausibility of a Theory of Everything (ToE) by applying principles from Gödel's Incompleteness Theorem.

From here, we will consider women of science and women of faith, reflecting on the need to elevate the dignity of women in our world and how the exploration of faith and science should be a field that seeks to encourage and embrace the role of women.

Next, we will explore the delicate weaves of science and faith, such as how the language of "spirit matter" that comes from Teilhard de Chardin can be viewed through the language of Liturgy and Sacrament in the writings of Anglican theologian Catherine Pickstock and Pope Emeritus Benedict XVI.

We will also peer through the lens of G.K. Chesterton and of Flannery O'Connor to discover how literature, imagination, and science can come together in an honest, raw view of creation that deepens the truth of the world in which we live. Central to this honest view of creation is our smallness in the material world, discussed profoundly by Carl Sagan. Through Sagan's "Pale Blue Dot" reflection, we will explore the question of whether our smallness in creation leads to our significance or insignificance. Lastly, we will explore the mystical ascent of Saint Bonaventure and the genius of theological assimilation found in Saint Thomas Aquinas, exploring where the proper boundaries lie in an exploration of faith and science.

GEORGES LEMAITRE

FATHER OF THE "BIG BANG"

O NE OF THE BASIC questions of science has a rather surprising answer: Who was the first scientist to put forward the Big Bang Theory? Most would presume that it was Albert Einstein or Edwin Hubble. The correct answer is a diocesan priest from Belgium by the name of Monsignor Georges Lemaitre.

Lemaitre began his academic career at Louvain's College of Engineering in 1913. Due to World War I, Lemaitre was forced to leave his studies to serve in the Belgium artillery. After his military service, he entered the seminary, studying to be a priest for the Archdiocese of Malines. In his spare time as a seminarian, Lemaitre pursued his interests in math and science. After his ordination in 1923, Lemaitre was sent to study math and science at Cambridge where Arthur Eddington was the director of the campus observatory.

Lemaitre's first great academic work of science focused on

Einstein's Theory of Relativity. Lemaitre developed a hypothesis of an expanding universe that was later known as "Hubble's Law." The prevailing thought of the time was that the universe was eternal and static. Lemaitre disagreed and later put forth the idea of "The Primeval Atom," arguing that if the universe is expanding, it must have been compressed into a small and dense state in the past.

These hypotheses of Lemaitre set the groundwork for what would later be called "The Big Bang Theory." Lemaitre never used this term to describe his ideas. The first use of this term was by Sir Fred Hoyle during his popular series of radio lectures on "The Nature of the Universe" from 1950. Although he coined the phrase, Hoyle initially rejected Lemaitre's ideas.

A popular narrative often told of Lemaitre, one that may contain a co-mingling of fact and legend, is that his hypotheses were initially dismissed by the scientific community. It is said that Einstein's initial assessment of Lemaitre's hypotheses was that the calculations were quite good, but Lemaitre's grasp of physics was abominable. Over time, the harsh critique of Lemaitre's hypotheses changed as Edwin Hubble began to observe the red shift in the light spectrum of celestial objects, confirming that the universe is in a state of expansion. Because of these and other insights, Einstein retracted his earlier opinion on Lemaitre's work, calling it "the most beautiful and satisfactory explanation of creation to which I (Einstein) have ever listened."* In light of this, Lemaitre's ideas were vindicated and are foundational to the understanding of the Big Bang Theory we have today.

Lemaitre's brilliance was not only affirmed by the scientific community; it was celebrated by Pope Pius XII. In an age when the cultural presumption is that popes are looking to condemn scientists and reject their theories, the relationship between Lemaitre and Pius XII was quite different, showing this narrative of divisiveness to be in error.

In fact, it wasn't the Pope who questioned Lemaitre's theory of the expansion of the universe, it was Lematire who warned the Pope not to use the theory as a "proof" of Biblical creation. Lemaitre, as a good scientist, knew that with time his theory would be improved upon, shown to have errors, and/or be disproven all together. Nevertheless, Pope Pius XII embraced the work of Lemaitre and reflected on the science of the day in his Address to the Pontifical Academy of Sciences on November 22, 1951. In regard to the expansion of the universe, Pius XII said the following:

> The examination of various spiral nebulae, especially as carried out by Edwin W. Hubble at the Mount Wilson Observatory, has led to the significant conclusion, presented with all due reservations, that these distant systems of galaxies tend to move away from one another with such velocity that, in the space of 1,300 million years, the distance between such spiral nebulae is doubled. If we look back into the past at the time required for this process of the "expanding universe," it follows that, from one to ten billion years ago, the matter of the spiral nebulae was compressed into a relatively restricted space, at the time the cosmic processes had their beginning. (Address to the Pontifical Academy of Sciences on November 22, 1951, paragraph 36.)

Even in this statement, we can see Lemaitre's warning proven to be correct as science now thinks the initial "Big Bang" occurred some 13.7 billion years ago instead of the 1 to 10 billion years cited by Pope Pius XII. Thank goodness that Papal Infallibility only applies to faith and morals and not the natural sciences! The mere fact that Pius XII was willing, as Pope, to affirm the best science of his time sets a clear precedence for today that Catholics, and all people of good will, can trust scientific investigation, presuming, of course, it is done in a truly scientific manner.

The importance of people like Lemaitre is to remind both the scientific world and the Christian world that the Church supports true science. This does not mean that we must accept every scientific finding with the doctrinal weight of Church teaching. However, it does mean that Christians should avoid the petty wars that break out between faith and science, affirming that both are partners of dialogue in the exploration of truth, not treating each other like ideological punching bags.

* Much of the historical and biographical information for this chapter was taken from: Mark Midbon, *"A Day Without Yesterday,' Georges Lemaitre & the Big Band,"* Commonweal Magazine. Vol. 127 No. 6 (March 24, 2000) 18–19

STANLEY JAKI, OSB

THE PRIEST WHO QUESTIONED THE
PLAUSIBILITY OF A THEORY OF EVERYTHING

THEOLOGY MEASURES NOTHING, WHILE "exact science" deals only with numbers and measurements of material change. This is the core thesis of much of Father Stanley Jaki's approach to faith and science. Arguing that each discipline should strictly adhere to its own principles, Jaki strongly emphasized that the unique focus science has on the material world makes it impossible to create a "theology-science" or "philosophy-science." When reading and listening to Jaki's brilliant reflections, it becomes clear that his detailed critique on matters of faith and science illuminate the fundamental distinction that science deals with the "how" aspect of creation while philosophy and theology deal with the "why."

To demonstrate this distinction between faith and science, Jaki would often share a story about the properties of electricity. The story is of a young scientist who gave a factory tour to Lord Kelvin (1824–1907), arguably one of the greatest scientists of his time. The factory created equipment that measured the effects of electricity and was built by Lord Kelvin himself. Unfortunately, the young man giving the tour was not aware that the man before him was Lord Kelvin. After the young man spoke in great detail of how these gadgets measured electricity, Lord Kelvin thanked him for the tour but wanted to ask one last question, "What is electricity?" When the young man was unable to answer this question, Lord Kelvin consoled him by saying that he himself was equally ignorant of the answer. The moral of the story is that it is one thing to measure how electricity behaves, but it's a completely different thing to understand what electricity is at its essence. Jaki would use this story to argue that science and theology should not be combined, but rather they should stay within the parameters to which each naturally adhere.

We understand a thing's essence by looking at its actions tho

Fr. Jaki was born August 17, 1924, in Hungary and grew up to join the Benedictine Order. His academic background is most impressive. He received doctorates from the Pontifical Institute of Sant' Anselmo in theology and from Fordham University in physics under the tutelage of Victor Hess. He completed post-doctoral research in the philosophy of science at Stanford, Berkeley, and Princeton, and was an honorary member of the Pontifical Academy of Science, receiving the Templeton Prize in 1987. Before his death in 2009 of a heart attack, Jaki was a distinguished professor of physics at Seaton Hall University.

Regarding his contribution to science, Jaki is best known for his affirmation of Gödel's Incompleteness Theorem in light of modern physics search for a "Theory of Everything" (ToE). The theorem is rather complex to explain, but the main thrust is that it affirms that no mathematical theory can be completely self-sufficient and there will always be parts of a mathematical system that are either self-contradictory or unable to be verified. The significance of this to Jaki was that it promised a disappointing road ahead for scientists looking for a ToE that would be able to explain a world that is self-sufficient, meaning that its existence has no need of contingency upon a Creator. Jaki observed that in light of Gödel's Incompleteness Theorem a complete ToE was impossible. Here in his own words, Jaki reflects on Gödel's incompleteness theorem.

> Ideology seems to have played an important role in the resistance by prominent physicists to perhaps the greatest discovery in the history of mathematical logic, or Kurt Gödel's formulation, in November 1930, of the theorem that any nontrivial set of arithmetic propositions has a built-in incompleteness. The incompleteness consists in the fact no such set can have its proof of consistency within itself. The bearing of that incompleteness on physical theory, which has to be heavily mathematical, should seem obvious. (Stanley Jaki, On a Study About Gödel's Incompleteness Theorem)

Some mistakenly see this theorem as the "death of modern physics." But for Jaki, Gödel's Incompleteness Theorem is somewhat of a beginning, promising that science will never come to an end of exploration, but will always have new discoveries and advancements to explore. Stephen Hawking affirmed this sentiment in a presentation he gave on Gödel's theorem titled, "Gödel and the End of Physics."

> Some people will be very disappointed if there is not an ultimate theory that can be formulated as a finite number of principles. I used to belong to that camp, but I have changed my mind. I'm now glad that our search for understanding will never come to an end, and that we will always have the challenge of new discovery. Without it, we would stagnate. Godel's theorem ensured there would always be a job for mathematicians. I think M theory will do the same for physicists. I'm sure Dirac would have approved. (Hawking, "Gödel and the End of Physics")

This brief reflection on the thoughts of Jaki doesn't even scratch the surface of this great priest and scientist. However, it reminds me that an honest assessment of our limitations doesn't bring us to an end of our understanding of God and the world, but opens us up to new possibilities, promising an endless well of truth to draw water from. Whether our interests are in faith, science, or both, may we constantly be open to explore truth in our lives, embracing the never-ending pilgrimage that leads us to the God who is the source of all truth.

WOMEN OF SCIENCE,

WOMEN OF FAITH

REFLECTING ON INFLUENTIAL
WOMEN OF FAITH AND SCIENCE

As I HAVE GROWN in my appreciation of science, there are key people I've known who have both assisted and inspired me to embrace the study of the world around us. In a field that is made up primarily of men, I deeply appreciate the influence that brilliant women of science have had upon my intellectual growth. There are many I could reference, but three stand out in particular. The first is Dr. Anne Geraghty, a former colleague of mine when I taught at Regis High School in Eau Claire, WI. Geraghty's specialty was molecular biology. Our discussions on matters of faith and science helped me see clearly the problems with Intelligent Design and helped me rediscover the intellectual tradition of Catholicism that embraces science on its own terms.

The second is Dr. Brenda Frye, professor of physics at the University of Arizona and astronomer at the university's Steward Observatory in Tucson. When I attended the first Faith and Astronomy Workshop, I enjoyed Frye's presentation on gravitational telescoping, including images from the Hubble Space Telescope. Gravitational telescoping involves gravity bending the light of distant celestial objects and reflecting their image in different parts of the night sky, at different stages of the object's evolution. It is often said that the night sky is like a history book, presenting the heavens not as they are, but as they were depending on their distance from us. Gravitational telescoping provides a history book within a history book as scientists study the life of a galaxy in new ways.

Lastly, I have been inspired by an up-and-coming woman of science, one of my parishioners at St. Joseph Parish who is a student at Massachusetts Institute of Technology. As I pray for her and listen to her dreams about being part of the exploration of objects beyond Pluto, I can't help but wonder whom she will inspire someday.

Similar to my appreciation of women of science, I can also point to many women who have helped me to understand the Church both as a priest and as a layman. In an environment of male only priesthood, my years of ministry have affirmed that collaboration with faith-filled women is essential in the life of the Church so that the full breadth of the Mystical Body of Christ can be expressed. Whether it be universal figures such as social justice activist Dorothy Day and Mother Teresa of Calcutta or women I have worked with in my daily ministry, I marvel at the examples of faith they have been and how they have shaped my approach to ministry.

I have also been blessed with friends, parishioners, and students who have invited me to join them in their walk with Jesus Christ. When faith is found and shared, it not only helps the person I walk with, it helps me. For example, one of my former students, Megan (Geraghty) Lobos, wrote an article for the United States Conference of Catholic Bishops on the challenges of international relationships. When I read this piece for the first time, my mind went back to conversations Megan and I had about her journey with her boyfriend Juan. These memories provided not only a moment of nostalgic pride, but gratitude on how their journey shaped my approach to priestly ministry.

Experiences like these have led me to see the lay people I work with as extensions of my priestly ministry. Without their work and dedication, I would not be able to accomplish that which God has called me to do: Care for God's people and invite them into a life of holiness.

These reflections remind me of Papal calls to uphold the dignity of women in the Church and the modern world. St. John Paul II, in his World Day of Peace Address in 1995, reflected on how a world that accepts and promotes the contributions of women improves society and promotes the cause of peace.

When women are able fully to share their gifts with the whole community, the very way in which society understands and organizes itself is improved, and comes to reflect in a better way the substantial unity of the human family. Here we see the most important condition for the consolidation of authentic peace. The growing presence of women in social, economic and political life at the local, national and international levels is thus a very positive development. Women have a full right to become actively involved in all areas of public life, and this right must be affirmed and guaranteed, also, where necessary, through appropriate legislation. (John Paul II, World Day of Peace – Women: Teachers of Peace)

Pope Francis has shared similar sentiments about the indispensable role of women in the Church, stressing the importance of promoting their dignity globally and of finding new ways for women to be involved in Church leadership. Often, these calls spur great optimism for the Church's future, but they can also lead to deep frustration, feelings that the Church is not doing enough for women. In either case, it becomes clear that the dignity of women in the Church is an ongoing pastoral need to address regardless of country of origin, economic status, or state of life.

In 2016, the Vatican hosted an event titled "Voices of Faith," encouraging women to share how their leadership helped strengthen the Church. Could there be an opportunity to invite and encourage women in the sciences to reflect theologically on their lives as Catholics and as scientists?

Four nuns whose names until recently were lost to history are silent witnesses to the legacy of Catholic women of science. Remembered only with simple photographs, these four women helped the Vatican Observatory catalog more than 480,000 stars. Their identities have been recovered so history can now know their identities. They are Sisters Emilia Ponzoni, Regina Colombo, Concetta Finardi, and Luigia Pinceri. Long before these women Religious assisted the Vatican Observatory, other women of faith and science broke monumental barriers for their time.

A woman of great influence who is often identified as a Catholic woman of science is Saint Hildegard von Bingen, who lived in the 11th and 12th centuries. Saint Hildegard was identified as a polymath, meaning someone who has a broad spectrum of intellectual expertise, displaying brilliance in many fields of study. She was known as a mystic, astronomer, philosopher, physician, and musician. Although she developed a theory of the origin and structure of the universe, it was more of a mystical vision based on the medieval worldview of her day and her periodic visions. Still, we can affirm and uphold her brilliance. Pope Emeritus Benedict XVI offered a beautiful reflection on the life of Saint Hildegard when she was declared a Doctor of the Church. Here is a section of that reflection, focusing upon her view of creation and the Trinity.

Hildegard asks herself and us the fundamental question, whether it is possible to know God: This is theology's principal task. Her answer is completely positive: through faith, as through a door, the human person is able to approach this knowledge. God, however, always retains his veil of mystery and incomprehensibility. He makes himself understandable in creation but, creation itself is not fully understood when detached from God. Indeed, nature considered in itself provides only pieces of information which often become an occasion for error and abuse. Faith, therefore, is also necessary in the natural cognitive process, for otherwise knowledge would remain limited, unsatisfactory and misleading.

Creation is an act of love by which the world can emerge from nothingness. Hence, through the whole range of creatures, divine love flows as a river. Of all creatures God loves man in a special way and confers upon him an extraordinary dignity, giving him that glory which the rebellious angels lost. The human race may thus be counted as the tenth choir of the angelic hierarchy. Indeed human beings are able to know God in himself, that is, his one nature in the Trinity of Persons. Hildegard approached the mystery of the Blessed Trinity along the lines proposed by Saint Augustine. By analogy with his own structure as a rational being, man is able to have an image at least of the inner life of God. Nevertheless, it is solely in the economy of the Incarnation and human life of the Son of God that this mystery becomes accessible to human faith and knowledge. The holy and ineffable Trinity in supreme Unity was hidden from those in the service of the ancient law. But in the new law of grace it was revealed to all who had been freed from slavery. The Trinity was revealed in a special way in the Cross of the Son. (Pope Emeritus Benedict XVI, Apostolic Letter proclaiming Hildegard of Bingen, professed Nun of the Order of Saint Benedict, a Doctor of the Universal Church. Section of paragraph 4)

Another interesting study is the role Pope Benedict XIV played during the enlightenment, leading to the promotion of women scientists and mathematicians. Laura Bassi (1711–1778) was the first woman to officially teach at a European University. What is even more amazing is that she began teaching at the University of Bologna at the age of 20. Bassi was known for her understanding of Newtonian physics, experimental physics, and electricity. Cardinal Prospero Lambertini, who later became Pope Benedict XIV, recognized Bassi's intellect and became one of her biggest supporters. The future Pope encouraged her not only to study the emerging science of the times, but to participate in public lectures as a symbol for the City of Bologna. Bassi was eventually appointed to the Pope's group of 25 intellectual elites, called *The Benedettini,* whose goal it was to stimulate new scientific research in Bologna. Because she broke barriers for women in professional education and the sciences, Laura Bassi is a historic figure not only for the Catholic Church, but for all people.

If we, as clergy, are to take seriously Pope Francis' call to elevate the dignity of women, we need to be attentive to the gifts God has given to those around us and actively encourage those gifts to be explored and shared for the good of the Church and the world. Tensions will remain around certain questions, like priestly ordination, but that should not and cannot be used as justification to exclude women from participation in leadership roles in the Church where it is possible. The Scriptural grounding for this comes from St. Paul, describing the Mystical Body of Christ as analogous to the human body. Each part of the body is unique and distinct, but, in order for the body to act as it should, all the members must act in concert as one.

Spiritual Exercise: Pray that God continues to call forth the gifts and talents of women to bless and enrich the Church. Together, may all of us, women and men, walk together on our common pilgrimage of faith. In this journey, may we value the voices of all who God has inspired to help build up the Church, enlivened by the diversity and unity of the Mystical Body of Christ.

TEILHARD DE CHARDIN AND

CATHERINE PICKSTOCK

UNDERSTANDING THE LANGUAGE OF CREATION

WHAT WAS IN THE beginning? Depending upon your personal interests, there are numerous ways to answer this question. The astronomer might explore the inner workings of the "Big Bang," making the discovery of gravitational waves an exciting doorway, hoping to shed light on the moment the singularity ceased to be a singularity and expanded into the known universe.

The theologian's mind might turn to the prologue of the Gospel of John, reflecting on the poetic strophes that affirm that "In the beginning was the Word (Logos)." As many theologians have commented, one of the distinctions between the creation story of Genesis and other creation stories of the same genre is that most mythologies depict creation as acts of violence between warring gods, while

Genesis presents creation as a "Word act" of non-violence.

I am fascinated by both the scientist who is listening for insight into the moment of creation and the Christian who listens for the voice of Him who first spoke all things into existence and whose fecund Word continues to speak to this day. In both instances, there is a disposition of attentive listening for the "sound" of our genesis.

Now, I must be careful not to misrepresent the two "songs" of gravitational waves and the prologue of John as being one and the same. As we understand how scientists listen for gravitational waves and how Christians listen for God, there is a greater dissimilarity than similarity to these approaches when understanding God's creation. Despite this dissimilarity, I can't help but see in these parallel themes a hint of C.S. Lewis who depicts creation as a song in his fictional work, *The Magician's Nephew.*

> In the darkness something was happening at last. A voice had begun to sing. It was very far away and Digory found it hard to decide from what direction it was coming. Sometimes it seemed to come from all directions at once. Sometimes he almost thought it was coming out of the earth beneath them. Its lower notes were deep enough to be the voice of the earth herself. There were no words. There was hardly even a tune. But it was, beyond comparison, the most beautiful noise he had ever heard. It was so beautiful he could hardly bear it...

> Then two wonders happened at the same moment. One was that the voice was suddenly joined by other voices; more voices than you could possibly count. They were in harmony with it, but far higher up the scale: cold, tingling, silvery voices. The second wonder was that the blackness overhead, all at once, was blazing with stars. They didn't come out gently one by one, as they do on a summer evening. One moment there had been nothing but darkness; next moment a thousand, thousand points of light leaped out—single stars, constellations,

and planets, brighter and bigger than any in our world. There were no clouds. The new stars and the new voices began at exactly the same time. If you had seen and heard it, as Digory did, you would have felt quite certain that it was the stars themselves which were singing, and that it was the First Voice, the deep one, which had made them appear and made them sing....

The earth was of many colours: they were fresh, hot and vivid. They made you feel excited; until you saw the Singer himself, and then you forgot everything else. It was a Lion. Huge, shaggy, and bright, it stood facing the risen sun. Its mouth was wide open in song and it was about three hundred yards away. (C.S. Lewis, *The Magician's Nephew*.)

Creation as song is a beautiful image for someone like me who was first trained to be a musician. However, at a more basic level, we also need to appreciate the fact that we can reflect on such mysteries in the first place. Do we take time to understand how it is that we are able to be aware of ourselves, our neighbor, our world, worlds beyond our world, and, lastly, the deepest of all questions, where did all of this come from in the first place? Between the moment of our beginning and our current moment of becoming is the realization that, without our consciousness, none of this exploration would be possible.

Consciousness is a great mystery in and of itself. Our consciousness allows both the astronomer and the theologian to listen for the moments of our beginning. As mentioned earlier, gravitational waves produce a type of celestial music, which is able to be measured when events like collapsing black holes occur. This and other phenomena that are observed in space lead to the creation of equations and symbols to make sense of what is observed, developing a language for science to be handed on to others.

For theology, to speak of a creative Word or Logos that was in the beginning points to a First Utterance or First Expression that speaks things into existence in a way that cannot be fully expressed in human language, but is still documented as oral Tradition becomes written Tradition. In both instances, scientific and theological, our consciousness emerges as our inner wiring to take what is experienced in creation and express it in a manner that is both understandable to the individual and able to be shared with the world through a sense of community.

Consciousness was a central theme of exploration for French Jesuit and philosopher Teilhard de Chardin. In his understanding of "complexification," Chardin observes that as matter organizes and becomes more complex, it also displays various levels of consciousness. The more complex a thing becomes; the deeper levels of consciousness a thing displays. At the highest levels of complexification, Chardin argues that what is found is matter seeking to become spirit or "spirit-matter." Though Chardin argues against two distinct realms of matter and spirit, we can see that self-consciousness explores the material and spiritual aspects of creation in different ways.

Science has created a language that expresses what we can call a "natural consciousness," or the things that can be observed and measured. Religion, too, has a distinctive language that expresses what we can call a "meta-consciousness," or our ability to not only comprehend both creation and a Creator, but affirm that there is something in creation that allows us to share in the very life of the Creator. This distinctive language of religion is found in profound ways through acts of Worship.

Regarding the language of worship, Dr. Catherine Pickstock, Anglican theologian at the University of Cambridge, presents one of the more fascinating treatments of the "traditional Mass" (the Tridentine Rite) in her work, *After Writing: On the Liturgical Consummation of Philosophy*.

Arguing against the mentality that justified the reform of the traditional Mass from the standpoint of simplifying that which had become overly complex, Pickstock argues that the repetitive nature of the traditional Mass does not present a cumbersome linguistic stumbling block to authentic worship, but provides the worshiper with an experience of "liturgical stammer."

This "stammer" sees in the repetitive linguistic structure of prayer a moment of awe in the presence of that which is beyond us and creates a constant moment of "beginning again," or re-creation, through God's closeness to us in the Eucharist. Pickstock's view is an interesting parallel to Chardin in that "comflexification in liturgy" does not always mean that something needs to be changed or simplified in our act of Worship, but can actually give voice to the complex mystery of how our spirit-matter encounters the spirit-matter of Christ in the Eucharist.

Pope Emeritus Benedict XVI gave voice to this dynamic of Chardin's understanding of complexification and the Eucharist in a homily given at Vespers on July 24, 2009.

"Let Your Church offer herself to You as a living and holy sacrifice". This request, addressed to God, is made also to ourselves. It is a reference to two passages from the Letter to the Romans. We ourselves, with our whole being, must be adoration and sacrifice, and by transforming our world, give it back to God. The role of the priesthood is to consecrate the world so that it may become a living host, a liturgy: so that the liturgy may not be something alongside the reality of the world, but that the world itself shall become a living host, a liturgy. This is also the great vision of Teilhard de Chardin: in the end we shall achieve a true cosmic liturgy, where the cosmos becomes a living host. And let us pray the Lord to help us become priests in this sense, to aid in the transformation of the world, in adoration of God, beginning with ourselves. That our lives may speak of God, that our lives may be a true liturgy, an announcement of God, a door through which the distant God may become the present God, and a true giving of ourselves to God. (Pope Emeritus Benedict XVI, Homily at Vespers, July 24, 2009)

What was in the beginning? As long as the human person is able to contemplate the big questions of life, we will apply our senses to try and gain insight into our origins. What is amazing is that whether it is through the music of gravitational waves or the divine utterances of the Word made flesh, our consciousness points us to not only the things of the tangible world, but also the transcendent.

One of the great mysteries of creation is that the song that sung us into existence also put within the human person the ability to participate in the Divine Chorus of the Trinity. At the center of this Hymn of love is the Eucharist, the fruit of the Cosmic Liturgy, which is our source and summit, our food for eternal life, and that which brings about a true change within the recipient to become what we receive.

Spiritual Exercise: I would encourage you to pray this week to understand your beginning, your becoming, and how God is calling you to participate in His very life. Together, let us join our voices to the chorus of creation, seeking to understand our place in the universe.

G.K. CHESTERTON

SO A PIGMY GOES TO MASS IN A MULTIVERSE
AND EXPERIENCES A SEVERE CASE OF ANAMNESIS

IF YOU'RE FAMILIAR WITH G.K. Chesterton, cosmology, and liturgical theology, you may be thinking this title gives away the entire chapter, leaving no reason to read on. If you're not familiar with any of these subjects, the title may sound like nothing more than gibberish, again, leaving no motivation to read on.

In either case, I invite you to enter into this "collision of worlds" to experience what I was trying to communicate with this title: A brief moment of intellectual play while exploring the subject of faith and science. To do this, we will delve into a fairy tale from G.K. Chesterton titled *Tremendous Trifles*, explore a little astronomy and cosmology, and conclude with a touch of liturgical theology and see what comes of this symphony of ideas.

One of my former professors, Dr. David Fagerberg, has a deep passion for the writings of G.K. Chesterton. In his article, "Humility without Humiliation: A Capacitation for Life in Elfland in the Thought of G.K. Chesterton," Fagerberg presents a delightful stitching of the wit and wisdom of this colorful figure in Catholicism. To accomplish this, Fagerberg begins with Chesterton's fairy tale, *Tremendous Trifles*.

For those unfamiliar with this fairy tale, it's the story of two young boys named Paul and Peter. A passerby talks to the boys and promises he will grant them each one wish. Paul quickly says he always wanted to be a giant so that he could easily walk around the world, taking in the wonders like the Himalayas and Niagara Falls. Paul's wish was granted, and he became a massive giant, able to easily bound through all of creation.

The irony for Paul was that after he became a giant, the wonders of the world he so desired to experience didn't seem so wondrous after all. The Himalayas no longer appeared as towering mountains but as rocks in a garden. When he viewed Niagara Falls, it no longer thundered as raging waters but trickled like the water coming from a bathtub faucet. His giant disposition led him to boredom and finally his demise.

Peter, on the other hand, chose something quite different when using his wish. He asked to be what Chesterton calls "the pigmy," only about a half of an inch high. From this small disposition the world became even more wondrous to Peter. Unlike Paul who became bored from his perspective as a giant, Peter set out for endless adventures in his smallness, having not completed them to this very day. (*Tremendous Trifles, G.K. Chesterton*)

Fagerberg explains that these "little boys" signify two approaches to how we view the world. The boy, Paul, who wishes to be a giant, is out of proportion to the world, choosing pride over humility, and, in the process, loses his sense of wonder as he chases after all the wonders of the world. The boy, Peter, on the other hand, chooses to be small, the pigmy, embracing a stance of humility toward the world. The end result for Peter is an endless sea of wonder from his small state.

Chesterton uses this story to argue that the sin of Satan is to view creation as something tiny and minuscule that can be controlled and dominated. Instead, the Christian disposition, the disposition Chesterton claims as his own, is that of the pigmy: a radically small creature that seems to have little if any significance. From this small vantage point, the pigmy lives in constant awe and wonder at the marvels of creation, leading to one of the most often quoted lines of Chesterton, "The world will never starve for want of wonders; but only for want of wonder." (*Tremendous Trifles*, G.K. Chesterton)

One may question referring to a fairy tale in a book on faith and science, but what modern astronomy teaches us is that the moral of this fairy tale is truer than even Chesterton may have intended. When gazing upon the vast expanse of the known universe, we are the pigmy, small and insignificant, looking in wonder at a universe that we cannot fully comprehend. Although the human heart deeply desires to become the giant that can leap from our Milky Way Galaxy, to the Andromeda Galaxy, to M51 like a child leaping on top of large stones in a shallow river, reality shows us that Peter's response to be "about half an inch high" is our material fate. So with no magic potion at our disposal, here we are, the tiny pigmy, looking up at the vast darkness of the night sky, in constant wonder over a simple question: What is out there?

As we gaze at the night sky in our smallness, we have come to learn some fascinating truths about our universe. One of them being that we see the objects of the night sky in relation to the distance their light had to travel to reach us. There are many helpful analogies we can use to understand this phenomenon. While in college, I was first introduced to the idea of the night sky as a history book, peering back, chapter by chapter, at time's origin.

Another helpful analogy was presented by a fellow contributor to the *The Catholic Astronomer*, Dr. Brenda Frye, in her post "Looking back in time…" in which an astronomer is like an archeologist, digging back further and further into time, receiving a glimpse into our past with the discovery of each galactic "fossil." What I find amazing about the metaphor of both historian and archeologist is that the astronomer is dealing with "living history." When people gaze at a night sky object, it is like they are watching an old video of Babe Ruth at the plate before he calls his shot, but seeing it as if it happened for the first time.

If the night sky were like a history book about Napoleon, we would experience far more than a mental reconstruction of wars told through words. Rather, we would see Napoleon standing tall (as tall as he could) to wage war as if we were viewing the battlefield from afar (or frighteningly close). Therefore, we are not dealing with a history or archeology that is a musty book on a shelf or boney remnants of a past species. Rather, it is a living history, being made present to us again, even if some of the objects we gaze on no longer exist.

This concept leads to a fascinating question: What if there is a real living history of the universe? Is there a way for us to manipulate the fabric of time so we could jump on the stones in the pond, not as a giant, but as a pigmy? What if our smallness could become so small that we could survive entering a worm hole to be transported to a part of the galaxy that, as of now, is impossible to reach? What if we now live in a multiverse in which each universe is like a droplet of rain on the hood of my car after a storm? What if there are parallel universes where other "mes" exist at the same time? Is this not the grandest of all cosmological fairy tales?

Doesn't the non-scientist (and perhaps some scientists) hear these reflections and wonder what episode of Star Trek or Dr. Who I am referencing? And isn't it also true that, just as Chesterton meant his story of the pigmy to illustrate the real importance of a humble disposition of heart in the real world, the astronomer and cosmologist who pursue these theories do so in the hope of discovering the truth of the material world? Just as well written fairy tales and science fiction can teach us much about the human condition, so, too, does good theology and good science assist us in the ascent toward truth.

Before people of faith dismiss these scientific investigations as science fiction, we only need to explore our own theology to hear faint echoes of a collision of worlds and realities: It's called the Mass. The theology of the liturgy is that we "reenter" the great events of salvation history in the timelessness of the Eucharist. This sense of timelessness is not meant to be a remembering of our past like the pictures I hang on my wall. Rather, the Greek word used to explain this reality is anamnesis. It is, for theology, a unique remembrance in which we are present once again at the Incarnation of Christ, his life, death, and resurrection. We are present at the exodus of the Jewish people, the Pentecost of the Holy Spirit, and the future "New Heavens and New Earth" referenced in Revelation.

This theology of anamnesis is derived from the words of institution at Mass, "Do this in remembrance of me." The English word "remembrance" is where we find the Greek word *anamnesis*, re-entering the one, eternal sacrifice of Jesus Christ. Therefore, this unique understanding of remembrance is rooted in the Eucharist itself, allowing the pigmy to skip upon the stones of the lake of salvation history, while still maintaining a sense of wonder and awe we call in theology, "the fear of the Lord."

At this point, the hard cynic may protest that this idea of anamnesis cannot be true because when we attend Mass we sit in pews, not in a first century Jerusalem home. Do we actually see Jesus on the cross again? Do we hear God say, "Let there be light?" Do we see Jesus in the upper room asking us to examine his wounds?

I could turn to the cynic and ask, "Have you ever seen a multiverse or a parallel universe," to which I might receive pages of equations that show the mathematical probability of their existence. The cynic may say science has gotten to the point that if a thing can be shown to exist in mathematics, it most likely exists in reality, challenging me to show any proof that anamnesis is real.

To that, I would say that I experience anamnesis every time I walk into a hospital room to anoint someone who is near death, gasping for every breath, with family members keeping watch at the foot of their loved one's "cross." In that moment, I am made present again to the Cross of Jesus Christ. When parents ask me to baptize their child and I allow water to flow upon the crown of the infant's head, I am present again at the moment of creation in which we were all, to quote Chesterton, "younger than sin." And when I see a parishioner return to Mass for the first time after a severe brain surgery and we embrace with joy because his greatest fear of irreversible memory loss miraculously has not occurred, I am present in the upper room once again to the wounds of Jesus Christ, hidden beneath the baseball cap upon his shaved head.

At this point, I will not turn around and become the giant, taunting the cynic. Rather, I will respect the cynic and simply say that science has yet to verify what the math suggests. What saddens me is that what can be verified of anamnesis among Catholics has, in turn, been reduced to a series of theological "equations," distant from the lived reality of the faith of many. It is time to allow our own "math" to point us to a reality that we ignore. We, as Christians, need to acknowledge that we have chosen the heart of the giant that ceases to find awe and wonder in the world, wanting our faith to be reduced to a Mass that is forty-five minutes long and a faith that conforms to our desires. Let us remember, as science reminds us, that we are the pigmy. Let us put down our endless chasing after material wonders, and in that detachment rediscover our sense of wonder at the timeless love of Jesus Christ.

Perhaps faith and science have more in common than we first thought. Perhaps our problem is that we have treated each other like two bookstore junkies engrossed in their own interests, forgetting that other sections of the store are waiting to be explored. Perhaps we need to acknowledge that we both, at times, have acted like the giant, needing to re-embrace the heart of the pigmy. Let all of us gaze in wonder at the world we live in and may all of us enter into the timeless wonder known as the love of God.

FLANNERY O'CONNOR

FAITH, IMAGINATION, CREATION, AND CO-CREATION

W HEN I BEGAN TO study the art of composing music, my professor shared that musical works that survived the test of time usually focus on one of three things: love, nature, or God. There is a characteristic of both the music and the message that resonates so deeply within our human nature that these works have been given the designation "classics." These classics remind us that each person yearns to express an inner desire for love and for a connection with both the natural world and the Divine.

As a simple test of this theory, let's say you just started a new job. When you're taken to your office or cubical, you immediately notice that there are no windows, no pictures, and nothing warm or living in your new beige box. After about a week of orientation and experiencing a couple "nutty" moments while looking at work binders and beige dividers, you decide to "liven up" your work space. Let me ask, with what do you choose to personalize your office? It wouldn't surprise me if you would put up pictures of your family (people you love), a potted plant or painting of the outdoors (nature), or perhaps some religious symbols or inspirational quotes (your faith). Even the things we surround ourselves with often point to love, nature, and God.

The human heart not only desires to gaze upon these things from the outside in, but also wants to experience them from the inside out. What I mean is that we not only want to hear songs about love, we want to be loved. We not only want to look at a beautiful painting of nature, we want to be a part of nature. And we not only want to read about God, we want to be in contact with God to help our lives find purpose and meaning. This two-fold desire is the foundation of participation in the theological sense: As we participate in the natural world we live in, and as we encounter authentic expressions of love with one another (in all of its manifestations), we are drawn into participation in the very life of God, seeking to be fully alive in Christ.

This theology and spirituality of participation breeds within the human heart a desire to create and co-create. When a young husband and wife share their gift of sexuality, it is not only an act of love with one another, but also an expression of their desire to co-create, to bring life into the world, intuitively affirming the beautiful sentiment from the *Catechism of the Catholic Church* that children are the "crowning glory" of marriage.

When artists wish to express on canvas the inner reality of a scene from nature, they do so, in part, to allow their creativity to become a part of creation. And just as a skilled artist has a discernible style of expression that becomes self-evident to the beholder, so, too, do we find the fingerprints of God upon creation, one of which being the desire to create in imitation of the Source of creation. Put another way, the Creator creates so that creation, in turn, can participate in the very life of God as co-creators.

This principle we're exploring can be a bit abstract, but it affirms that the human experience is not merely one of survival. Rather, we have within ourselves the desire for meaning and purpose, a desire not measurable with tools made by human hands. Where does this desire come from? Does this not point to a spiritual faculty of human existence that is unique in creation? And isn't this desire to be creator and co-creator so strong that to deny this inclination is to deny a fundamental part of who we are?

Some of the most horrific moments of human history, such as the Holocaust, have as a central sin the dehumanization of the person, forcefully denying each individual's dignity as a child of God and treating people as nothing more than a virus that needs to be eradicated or as wild animals to be caged and exterminated. These acts of horror remind us that when we lose the transcendent quality of life, we enter a "post-human" age in which life is defined by usefulness and productivity in contrast to being an expression of God's love in the world whose very existence invites them to participate in creation, whatever their ability may be.

Perhaps you would argue that Christianity actually limits authentic creativity. Some may say that it is only when one breaks the shackles of Christian influence that creativity is sincere, divorcing one's self from Catholicism's fixed pallet of symbols and meaning. A central problem with this protest is that it ignores authentic Christian anthropology. The protest presumes a "deductive" process of forcing a view of self upon the individual from outside.

The Christian understanding of human nature is "inductive," meaning God has placed within our being a longing for truth, goodness, and beauty, rooted in being made in God's image and likeness, though marred by sin. Therefore, the Christian life is not one of imposing a view of self from the outside, but rather to allow the true self to emerge that is from within. At the core of this struggle, however, is the humility to explore the depths of this inner self, both its beauty and its brokenness, so that, healed by God's grace, the true self becomes more self-evident.

It is this lack of humility among Christians that can lead some to argue that the Church seeks to stifle the soul, not letting the true self to be revealed. However, it is not the Church that stifles the creativity of the person, but, rather, it is the person who stifles themselves when refusing the humble exploration of who they are in God's eyes, despite the darkness they may encounter.

A way to understand this process in literature can be found in the work of Flannery O'Connor. In an essay she wrote for *America Magazine* on the relationship between fiction writers and Catholic dogma, Flannery argues that removing faith from the worldview of a novel is not creation by addition, but rather creation by subtraction. The novelist who presumes that a work can only be authentic if it excludes faith limits the work by excluding an essential aspect of human existence.

Yet Flannery also argues that when the faith of someone is weak, it is then they will recoil at the idea of writing a fictional novel that presents an honest, raw depiction of reality. It's important to note that Flannery is not making an argument for the writing of a tidied up, pristine book, ignoring the struggles of fallen human nature. Rather, the Catholic author should embrace writing on the honest, hard reality of human experience, which includes the movements of God's grace. To omit this grace is to omit a key dimension of human existence.

Whether we are gazing at the stars of the night sky, walking through the halls of an art gallery, enjoying the harmonies of a great symphony, or reading a novel about the struggles of life, we encounter a core desire to find love, to be connected with nature, and to experience both through the love of God. Take some time this week to pray for the people you love, to admire God's creation, and to allow your heart a moment of quiet contemplation in God's presence. In those moments, don't be surprised if you experience a stirring to add your own creativity to this vast creation in which we live. And when that moment comes, realize you have discovered one of the clear traits left within the human soul by God: To be co-creators in imitation of the Creator.

CARL SAGAN

PERSPECTIVE ON THE "PALE BLUE DOT"

ONE OF THE MOST impactful reflections on our place in the universe is Carl Sagan's "Pale Blue Dot." The anniversary of the iconic image of our earth as "a speck of dust" in our solar system was remembered a short time ago. Personally, I love Sagan's reflections on life and our small, blue home. Yet, amid this brilliant reflection, there is one part of the pale blue dot that provides a moment of discomfort: Sagan's bold statement that we are delusional for thinking we have a privileged place in this universe.

Now, from the standpoint of a material analysis of the world, Sagan is correct. When analyzing our material existence in relation to the rest of the universe, we make bacteria on a wet mound in a science lab look like the Milky Way galaxy. Yet, there is something that must be considered when exploring our place in the universe: We are able to understand just how small we are.

This realization reminds me of a pivotal experience from my college years. One night, while exploring the night sky with my 4-inch reflector, someone I knew well who was out biking dropped by to see what I was looking at. I was observing Jupiter and its moons. I let him look through my telescope and we began to talk about how Jupiter is a gas giant, a "potential star" that just didn't have enough "juice" to switch on. We talked about how Jupiter's moons are like a mini solar system of dynamic worlds— Io's sulfur volcanoes, the curiosity of what is under Europa's icy crust, and so forth. We then talked about astronomical units, light years, millions of light years, and billions of light years. At that point, the bicyclist gave an overwhelmed reaction of amazement, sparked simply by looking through the small eyepiece on my telescope. He said, "We are so small, so insignificant... we are nothing... how can you believe in God, knowing what you know about our universe, and think that somehow we are special in this universe?" My immediate response, "The fact that I can look through this telescope, realize how small we are, and how wondrous creation is strengthens my faith that there is a God who, for whatever reason, is allowing me to understand these things."

One of the most brilliant, short statements I have heard about our place in the universe comes from the former director of the Vatican Observatory, Fr. George Coyne, SJ. In a reflection he gave on the history of time and understanding it as a calendar year, he stated that the human person has been present on this calendar for two minutes and Jesus Christ for two seconds. At first this may seem to only affirm Carl Sagan's reflection that we are delusional if we think we are significant in this world. Yet, one of the fascinating things about the two-minute existence of the human person is that at this point of history, for some reason, creation is reflecting upon itself through us. This begs the question: Why?

As a Catholic priest, the exploration of this "why" has governed my whole path of life. The journey of understanding my place in the world, as small as it is, brought me to explore who the person of Jesus Christ is in my life. In that exploration, I find it fascinating that the Incarnation, the coming into our world of the Second Person of the Trinity, did not occur like a star going supernova or the massive explosion of a singularity that has expanded into the universe as we know it. Rather, it came as a "speck of dust" in our created world – hidden in the womb of a poor Jewish woman who would have been looked at with an eye of suspicion, given the circumstance of Jesus' conception.

On the night of Jesus' birth, a passerby might have heard Mary's cries from the cave and perhaps a "pale blue dot" was in the sky that seemed a little out of place. Would we have been surprised if the initial response to this scene would have been to ponder how insignificant this child and his family were? Yet, to reduce Jesus to the simple origins of where he was born would be to miss the great significance of who he is for the world: Lord and Savior.

My affirmation of Jesus Christ as Lord ironically brings me to the same conclusion that Carl Sagan came to as he reflected upon our material insignificance: We should treat each other with more kindness and gentleness, caring for one another and the planet we live on, because it is our home. Through the eyes of Christ, it is not the only home we will ever know, but it is the place we are called to be good stewards of and practice true charity toward, training us to become citizens of the Kingdom of Heaven.

Spiritual Exercise: How do you find significance in a world that can so easily make us feel insignificant? How do you discover a sense of meaning in a world that can so easily turn us into mere bacteria on a wet mound? I find my significance through gazing upon the beauty of the night sky and the beauty of my faith in Jesus Christ.

SAINT BONAVENTURE AND DR. MICHIO KAKU

HAS STRING THEORY PROVEN THE EXISTENCE OF GOD?

HOW DOES THE HUMAN person come to know God? This core question of life rests at the heart of many of my reflections for *The Catholic Astronomer*. One of Catholicism's foundational principles is that natural reason and Divine Revelation are the two wings on which the soul ascends to God. In this reflection on Saint Bonaventure, we will explore an understanding of spiritual ascent that is aided by six wings revealed through an intense, mystical experience. As we explore Saint Bonaventure's mysticism, we will come to see how Franciscan spirituality, greatly influenced by the thought of Saint Bonaventure, affirms the exploration of the natural world and how this exploration leads us to the knowledge of God. Second, I will compare Bonaventure's ascent with recent scientific speculations on the possibility of a Creator due to String Theory.

Saint Bonaventure was born in Bagnoregio, Italy, in 1217 and died in 1274. He joined the Franciscan Order in 1243 and studied at the University of Paris where he eventually was accepted as a Master of Theology. Bonaventure was known for his brilliance as a theologian and teacher while at the University of Paris, but was later removed from his position when he was named Minister General of the Franciscan Order.

There are many worthwhile topics to explore in Bonaventure's life, but the central event he is known for is a mystical experience at Monte La Verna. Monte La Verna is also known for Saint Francis of Assisi's vision, culminating in his reception of the stigmata. Consistent in both Bonaventure's and Francis' visions was a six-winged seraph. In Bonaventure's vision, the seraph became a mystical symbol in his written work, *The Mind's Road to God*, in which the Seraphic Doctor sees in the six wings of this heavenly vision a symbolic pathway of ascent for the human person when moving from the material world to God. Bonaventure explains that the six wings exist in three pairs, or categories, that are the exploration of the material (or sensible) world, our inner reflection on God's image and likeness, and, lastly, the revelation of the very essence of the Triune God as three persons in one nature.

With Bonaventure's exploration of the material world, we see a mix of philosophy and an approach to studying the natural world that is at home to the modern sciences. In regard to philosophy, Bonaventure speaks of understanding the material world as shadows and vestiges. This language rightly evokes the image of Plato's cave, needing to be freed from the shadows of our intellectual shackles and experience inner liberation by the light hidden from our sight beyond the cave.

Yet, to reduce Bonaventure's approach to the natural world as a theological representation of Plato alone would be a rash oversimplification. Bonaventure also affirms the thought of Aristotle and the exploration of truth through tangible things. The importance of this is that Bonaventure's language of shadows and vestiges is not a dualism that minimizes the importance of creation. Rather, Bonaventure draws upon both Plato and Aristotle (and aspects of neoplatonism) to affirm the importance of creation and the necessity of understanding the physical world as the beginning of the soul's ascent to God.*

To further understand creation, Bonaventure encourages the study of the sensible world in a way that is at home with the modern sciences: observe, weigh, measure, number, and compare. As we do these basic measurements, there emerges a self-evident order and beauty. The more we study a thing in itself, the more it exhibits qualities that point to something beyond itself, ultimately pointing to its Origin.

Key to understanding Bonaventure's model of ascent is our ability to internally process and reflect upon the order and beauty of things. As we understand and take delight in a thing we observe, we encounter the second set of wings of ascent, allowing us to discern God's image and likeness in this world of created things. In a real way, all things bear God's image, but only humanity bears both God's image and likeness. However, we should not read Bonaventure's distinction as purely academic, but as being intimately tied to his own mysticism and that of the founder of his order, Saint Francis. Pope Francis reflected upon this in his encyclical Laudato Si' when speaking of Saint Francis' mystical view of creation.

> (Saint) Francis helps us to see that an integral ecology calls for openness to categories which transcend the language of mathematics and biology, and take us to the heart of what it is to be human. Just as happens when we fall in love with someone, whenever he would gaze at the sun, the moon or the smallest of animals, he burst into song, drawing all other creatures into his praise. He communed with all creation, even preaching to the flowers, inviting them "to praise the Lord, just as if they were endowed with reason." His response to the world around him was so much more than intellectual appreciation or economic calculus, for to him each and every creature was a sister united to him by bonds of affection. That is why he felt

called to care for all that exists. His disciple Saint Bonaventure tells us that, "from a reflection on the primary source of all things, filled with even more abundant piety, he would call creatures, no matter how small, by the name of 'brother' or 'sister'." Such a conviction cannot be written off as naive romanticism, for it affects the choices which determine our behaviour. If we approach nature and the environment without this openness to awe and wonder, if we no longer speak the language of fraternity and beauty in our relationship with the world, our attitude will be that of masters, consumers, ruthless exploiters, unable to set limits on their immediate needs. By contrast, if we feel intimately united with all that exists, then sobriety and care will well up spontaneously. The poverty and austerity of Saint Francis were no mere veneer of asceticism, but something much more radical—a refusal to turn reality into an object simply to be used and controlled. (Pope Francis, Laudato Si'. 11)

As we read the words of Pope Francis, we begin to see the language of scientific examination and spiritual mysticism begin to blur with and separate from one another. To help us make this transition from the natural world to the supernatural, Saint Bonaventure makes a distinction between apprehension and comprehension.

In his series of lectures, *The Intersection of Science and Theology: Evolutional Theory and Creation*, Father James E. Salmon, SJ states that, for Saint Bonaventure, "to apprehend is to grasp something of reality, but not be able to encircle it or define it. To comprehend is to grasp and be able to encircle something or define it like a mathematical theorem." (Father James, Salmon, SJ, Ph.D. *The Intersection of Science and Theology: Evolutionary Theory and Creation*. Lecture 12)

This helpful distinction shows us that as we study the physical world, there are those shadows and vestiges that we can comprehend, meaning we can understand them completely, leading to a clear definition that encapsulates the thing being observed. However, as we reflect upon image and likeness, the things we comprehend begin to take on a new dimension that points to something beyond the thing itself, like a footprint or fingerprint of the Divine, which also unites all things. This type of recognition is what Saint Bonaventure means by the term "apprehend," meaning that we can grasp a thing's reality, but cannot fully encircle or define all of its aspects.

This distinction is not a mere recognition of gaps in scientific knowledge that will be closed someday. Rather, it is a humble recognition that scientific language has limits, able to occasionally point to the possibility of something beyond our natural world, but lacks, by its very nature, the ability to encircle or define that which is beyond nature. It is at this point that Saint Bonaventure introduces the final wings of ascent in the illumination of the soul, which is to understand the unique mystery of God's oneness of Nature and Trinity of Persons, revealed to us through Divine Revelation.

As I reflect on Saint Bonaventure's insights, I can't help but think of recent social media claims that can be summarized as, "Science has confirmed God's existence." As odd as this might sound, my first reaction as a priest to these bits of cyber click-bait is deep skepticism. From a pastoral perspective, I want to make sure the people I serve receive the truth of our faith, free from over sensationalized claims that can do more harm than good to a person's spirituality. On a personal level, I know the feeling of being duped by these claims, hoping that some monumental breakthrough has occurred in our understanding of the world. In reality, it was someone's misguided attempt to stir up people of faith.

One of these forms of cyber deception that is rather popular these days is the claim that scientist Dr. Michio Kaku has found definitive proof of God's existence. When digging into this, I came to two conclusions: Kaku himself would probably reject this claim and his actual reflection fits nicely as an example of the first step of Bonaventure's mystical ascent. Let's explore what I mean.

I clearly recall the first time my fellow seminarians and I explored Saint Thomas Aquinas' "proofs" for God's existence. In an attempt to stave off any misunderstandings that could occur in this exploration, our teacher told us, "Now, these are not 'proofs' in the modern sense of the word, but better understood as ways of understanding how a logical creation points to a logical Creator." The teacher also emphasized that Thomas is working through these proofs from the presumption that God exists. Therefore, the move for Thomas is not from no belief in God to belief in God, but rather, Thomas' proofs are logical demonstrations of how the order and beauty of creation, to borrow language from Saint Bonaventure, points to an origin to this order and beauty (a first unmoved mover, a first undesigned designer, etc.).

Far are we, in these demonstrations, from an understanding of the God of the Old and New Testaments, the saving love of Jesus Christ in our lives, and the foundational movements of the Holy Spirit that are central to Christian belief. Rather, the point of the proofs is to demonstrate the logic of belief in a Creator.

This lecture came to mind watching Kaku's *Big Think* video entitled, "Is God a Mathematician?" This is the video many websites have used to claim that science has definitely proven God's existence. In the video, Kaku nicely lays out the development of physics and mathematics from Newton to string theory. In conclusion, he states that perhaps God is a mathematician and the mind of God is discovered through the music of strings resonating through the numerous dimensions of space. At this point we can ask the question, Has Kaku proven that God exists because of the science of string theory?

To begin with, if it was the intention of Kaku to demonstrate a "proof" for God's existence (which I doubt was his motive), this brief video would not ascend to arguments laid out by Saint Thomas. His demonstration lacks the logical sequence demanded by philosophy and theology to argue for such a proof. Nevertheless, I am intrigued that Kaku is essentially pointing out the core thesis that goes back to my reflection on Bonaventure, that a logical, ordered, and beautiful universe points to something beyond this universe that is its ultimate Source.

From this standpoint, we can affirm a consistency between Kaku's presentation of string theory and the first step of Saint Bonaventure's ascent that observes, measures, weighs, and numbers the shadows and vestiges of creation, demonstrating a profound sense of order and beauty. However, Kaku's final conclusion stops the ascent at this point and concludes that a good candidate for the mind of God is the resonating strings of string theory.

My first reaction is to simply ask Kaku, "Where did the resonating strings come from?" Can we not affirm that the order and beauty that is found in string theory intuitively points to something beyond these resonating strings to the very grounding for the strings' existence in the first place? This curious final cadence in Kaku's reflection on God being a mathematician also begs another question: What is the difference between a "proof" of God's existence and answering the question, "Who is God?"

Revisiting my seminary lectures on Thomas' proofs, it's always important to remember, as people of faith, that the logical demonstration of the existence of God is one thing, but it is a completely different exploration to answer Jesus' question, "Who do you say that I am?" As I did a little more research on Kaku's thoughts on God, I was not surprised to find that the God Kaku argues for is the "God of Spinoza." And who is the "God of Spinoza" referenced by Kaku? In Kaku's words, this God is the God of order, beauty, and rationality. Again, we find ourselves back at the first stage of ascent in Saint Bonaventure's exploration of shadows and vestiges.

So, we can see that modern science is not inconsistent with Saint Bonaventure's beginning of our ascent to God. However, the materialist tendency of modern science finds great comfort in simply affirming a soft agnosticism on the question of God with no further exploration into the topic. This, however, should not be a shock to us nor offend the serious Christian.

Because it is the nature of science to remain neutral on questions of God, the most that science can affirm is a soft agnosticism or the mere plausibility of a God. Therefore, as Christians try to demonstrate the reasonable arguments for God's existence, we need to avoid an implicit reduction of our defense of God's existence to only a logical demonstration. We need to reaffirm, as a people of faith, that ascent of the human soul toward God requires inner reflection on how God's image is present in creation, how God's image and likeness resides within the human person, and how God has come to meet his people in the person of Jesus Christ, asking us to come and follow him so that the God of beauty and order can bring about beauty and order within our soul through God's grace and a life of doing God's will.

Put another way, string theory may someday demonstrate the plausibility of God in a way similar to Saint Thomas' "proofs" of God's existence, but these theories are only the first step in a lifelong exploration of who we are in relation to God and who God is in our lives.

Spiritual Exercise: Do we really open ourselves up to a spiritual ascent into God or do we limit ourselves, only seeking a Creator that is comfortable on our terms? As we pray for the courage to open our hearts to this exploration, let us give thanks that the modern sciences have continued to reveal to us the order and beauty of the created world. From that foundation, may we seek God's image and likeness in our lives, allowing God's presence to be known to us through the gift of God's illuminating grace.

* Much of the biographical and theological points of Saint Bonaventure are a summary from the article, *Saint Bonaventure from the Stanford Encyclopedia of Philosophy.*

THOMAS AQUINAS

DOES THE CATHOLIC CHURCH NEED SOMEONE TO ASSIMILATE SCIENCE INTO THEOLOGY?

CONSIDERING THE ARGUMENTS BETWEEN faith and science, the Church needs another Thomas Aquinas. To those who have a passive interest in theology, this sentiment may seem odd. After all, Thomas Aquinas lived in the 13th Century and many of his writings are intellectual cornerstones for the traditional understanding of God, the Sacraments, ethics, and morality. Besides, Thomas Aquinas' work was written well before the invention of the modern sciences. In light of this, one may question whether or not Thomas Aquinas would bring something useful to the modern issues of the relationship between faith and science.

For those who are well-versed in theology, this sentiment makes a great deal of sense. In our previous reflections on God and creation, we looked at how the Church Fathers explored the intimate connection between the created world and the Divine. This exploration not only used Sacred Scripture, but incorporated truth that can be found in intellectual traditions outside of the Bible. In particular, the early Church drew heavily on the philosophy of Plato, understanding the things of Earth as pointing to a higher, metaphysical form.

What made Thomas Aquinas so ground breaking was that he incorporated the thought of Aristotle as the philosophical underpinning to much of his theology. Aristotle, one of Plato's students, chose not to emphasize metaphysical forms, but instead sought to understand metaphysical truth "in the thing itself." For example, if you were trying to understand a tree, Plato would see in a tree a "shadow" (harkening to his great analogy of the "cave") of an ideal tree in which all trees point to. Aristotle, on the other hand, would seek to understand the essence of a tree through studying the tree itself. This difference in method led to the iconic image of Plato and Aristotle standing side by side in the fresco *The School of Athens* in which the older Plato is pointing up while the younger Aristotle is depicted as well "grounded" with a gesture indicating his more "earthy" philosophy.

Aquinas' philosophical shift was not only academic, but pastoral. He incorporated Aristotelian logic from Muslim scholars, such as Avicenna, Algazel, Averroes, Avicebron, and Maimonides, to help missionaries defend Christianity. The primary work that displays this ancient apologetic is Thomas Aquinas' *Summa Contra Gentiles*.

The genius of Aquinas was that he not only understood Aristotle in a pure sense, apart from the Muslim (and Jewish) scholars of the time, he also was able to demonstrate how Christian theology was consistent with the logic of Aristotle, strengthening the defense of Christianity with the very logic that was being used by those who opposed the faith.

A well-known application of Aristotelian logic in the theology of Thomas Aquinas was his use of matter and form in relation to the transformation of the bread and wine into the Body and Blood of Christ we call "transubstantiation." Aristotelian philosophy is in the background as Aquinas explains the complete change of the substance of the bread and wine into the Body and Blood of Christ while the appearance, or "accidents" to borrow from Aristotle, of bread and wine remain. Therefore, the Total Christ is present under the appearance of bread and wine. Another example would be Aristotle's understanding of motion as something "moves" from potentiality to actuality.

How does this apply to my original statement that some think that, in light of the tension between faith and science, we need another Thomas Aquinas in our time? Many people rush to the presumption that, just as Aquinas was able to develop a defense of Christianity through the same philosophical categories that were being used to attack Christianity, so, too, should someone in our time develop a defense of Christianity by using the very logic of science, showing that Christianity is able to assimilate the sciences in the way Aquinas assimilated Aristotle.

Although this may seem reasonable on the surface, there are some real problems with this line of thought. The primary challenge is that both philosophy and theology deal with transcendental elements such as meaning, purpose, goodness, and beauty, while science intentionally brackets these categories to focus solely on the measurable and empirical. This means there is an essential piece missing in the relationship between faith and science that makes assimilation in either direction difficult and potentially dangerous. In light of this, I am pessimistic that a Thomas Aquinas-like assimilation of faith and science will happen or is even possible.

So what, then, is the solution to our cultural tensions between faith and science? As I have argued, the best approach is to let science be science and theology be theology. In doing so, I find three themes that emerge that can accompany faith and science in their mutual desire for truth: humility, pilgrimage, and awe and wonder (or Fear of the Lord). It begins with humility, realizing we are part of something that is far greater than we can comprehend. This humility brings us to the 'humus" in Latin, or "to the Earth," in which we come to realize that we are small in the material sense and called to embrace smallness of heart in the spiritual life, allowing Christ to lift us up. This awareness calls us to pilgrimage, or exploration, to help us better understand the world we live in and the God we love. This journey changes us, helps us understand new possibilities, and opens us to new ways of thinking. At the end of this pilgrimage, we discover awe and wonder, contemplating the inescapable beauty of creation and the glorious "transcendent horizon" in which we stand upon the great mountain of faith, reaching one hand to this world and the other hand to God (borrowing an image from Karl Rahner).

What I find interesting about exploring humility, pilgrimage, and awe and wonder is that, at their end, they express what the theological tradition names Fear of the Lord—being reduced to awe and wonder before the glory of God. As the Psalmist states, "The fear of the LORD is the beginning of wisdom." (Psalm 111:10) This accompaniment between faith and science does not bring us to an end, but to a beginning. We find in this mutual exploration an understanding of reverence in the broad sense for the world that God has created. God then invites us to plumb deeper into the transcendental "ocean" of God's love and mercy.

Yes, sin creates difficulties in this exploration, necessitating the gift of grace so we may be strengthened on this pilgrimage. Yet the fact the world can be known shows that God wants the world to be known (borrowing a thought from Fr. Gabor, SJ). I find it fascinating that, despite the logical desire to discover a move of assimilation between faith and science in the intellectual tradition of the past, the healthiest relationship between these two great disciplines is to allow each to be itself, leading us to a common beginning, standing in awe and wonder of creation and Creator, being called to continue our life's pilgrimage as we encounter truth, goodness, and beauty.

* Much of the historical and theological points of St. Thomas Aquinas are a summary from the work *Basic Writings of St. Thomas Aquinas: Vol. I. God and the Oder of Creation.*

CONCLUSION

SECTION TWO

IN Section Two, we discovered how men and women of faith have made monumental contributions to the natural sciences. However, these contributions did not compel them to abandon their faith in the face of technical advancements. Instead, their faith was deepened. In our modern narrative that frequently seeks to paint the Church as constantly looking to condemn scientific discovery, we begin to see a different story emerge. Whether it be Pope Pius XII's desire to incorporate the Big Bang Theory into the Church's sacred memory or Pope Benedict XIV encouraging the brilliance of Laura Bassi to pursue the emerging sciences of the Enlightenment, we discover a Church that is open and encouraging of scientific investigation as a means of deepening our understanding of the world around us.

This investigation can and should be accessible to all people, through all intellectual disciplines. In the writing of G.K. Chesterton and the reflection on writing by Flannery O'Connor, we see that literature, even fiction, can allow for a more accurate understanding of the world we live in, deepening our faith through embracing an honest view of the world around us.

It is in this honest view of creation that we discover our smallness and are confronted with the question, illuminated by Carl Sagan, of what our true place is in this universe. Although a material view of our existence provides only a view of insignificance, the Spiritual ascent described by Saint Bonaventure shows that an exploration of the natural world can help us discover who we are as bearers of God's image and likeness. This exploration does not require a new assimilation of science and theology, similar to Thomas Aquinas' assimilation of Aristotle into the language of the Church. Rather, this assent of the soul points out that the journey of faith and science can be summarized in the categories of humility, pilgrimage, and awe and wonder or Fear of the Lord.

This Fear of the Lord brings us not to an end, but to a beginning, seeking not to be afraid of a false notion of God as a cosmic jackhammer that is ready to destroy us. Instead, we find a loving God. Our only fear is that we will separate ourselves from the most central love relationship of our lives.

In our next section, we will explore this language of love through the Church's prayer defined as the "Cosmic Liturgy."

INTRODUCTION

SECTION THREE

In this section, we will explore a spiritual view of creation that can be understood as the "Cosmic Liturgy." The Cosmic Liturgy is one of the most ancient views of the prayer of the Mass we have today. Starting with a foundation of viewing all of creation in a perpetual act of praising God, we see in the Cosmic Liturgy an incorporation of every aspect of life, all having an undercurrent of a prayerful connection with the Sacred.

This language fell out of use over time, but is still central to the reality we affirm as Catholics that in every Mass the Heavenly Liturgy and the Earthly Liturgy meet in the Eucharist. This encounter with Christ not only impacts the Mass itself, it transforms the way we see the passing of time, the different liturgical seasons, and the ultimate questions of life and death.

THE HYMN OF CREATION

SUN AND MOON, BLESS THE LORD

Psalm 19:1 THE HEAVENS declare the glory of God; the firmament proclaims the works of his hands.

What does it mean to say that creation is an act of liturgy? At first, this may seem a bit odd given our tradition of seeing acts of liturgy as the Mass or some other kind of structured prayer. However, if we allow our understanding of liturgy to be broadened, we can find in Scripture examples of what the Church Fathers called the "Cosmic Liturgy." A beautiful text that reflects on creation as an act of liturgy is the song of Shadrach, Meshach and Abednego from the fiery furnace. The song gives voice to a core belief that all of creation is giving praise to God. In beautiful, antiphonal strophes, we encounter the earth, the heavens, angelic powers, the sun, the moon, the rain, the dew, the snow, the day, the night, and every creature on earth blessing the Lord. (Daniel 3:50–90)

The litany affirms that every part of creation is glorifying God by simply being. In this light, we see creation as an act of liturgy, and acts of Liturgy as being intimately connected with creation. This worship of God is not only for this world, but includes the angels and heavenly hosts. Therefore, both material and non-material realms are in a perpetual act of worshiping God.

As it says in the prophet Malachi, "From the rising of the sun to its setting, my name is great among the nations; Incense offerings are made to my name everywhere, and a pure offering; For my name is great among the nations, says the LORD of hosts." (Malachi 1:11)

As you can see, astronomy can be more than just classifying a bluish looking star under the categories of O, B, or A. Astronomy can be an exercise of seeing a dynamic part of creation give praise to God. Observing a planetary nebula becomes more than just viewing the inevitable end of our sun. Rather, it echoes that verse of the "Hymn of Creation" in which a seed must fall and die in order for new life to come about.

In these metaphors, we are called to die to self to allow a "new creation" to emerge. Amid these observations of Cosmic Liturgy, we begin to see that we are invited to participate in this praise of God as well, adding a small, humble, but necessary voice to the vast chorus of creation.

Spiritual Exercise: Rest in a park, the woods, or someplace where you can quiet your heart. Read through the Book of Daniel and look around you. What are the various parts of creation that are present to you? Reflect on how they participate in the praise of God by simply being. In that space, add your voice of praise to God and give thanks for being a part of this act of liturgy.

ADVENT

AMID CREATION'S GROANING, THERE IS HOPE

Advent is the Church's New Year. It's accompanied by a change in the Gospel that we read and the colors we use to adorn our worship space, calling us to joyfully anticipate the second coming of Jesus Christ in final glory before shifting our focus to the birth of our Savior at Christmas.

Although the secular idea of a New Year evokes images of celebration, the Church's New Year begins with a more pointed message of spiritual attentiveness, watching for the Lord's return. Yes, there is an undercurrent of the end times in this season, but the historical development of Advent also calls us to invite Christ into our lives on a daily basis, dispelling the darkness of the world we live in with the light of the risen Lord.

This spirituality of attentiveness and invitation is best summarized through the symbolism of the Advent wreath. The wreath's origin is found in Scandinavian culture where creating a wheel-like wreath of evergreens encircled with candles marked the passing of winter. Day after day, a candle would be lit to anticipate the time when the days would begin to grow longer, celebrating the return of sunnier days after the winter solstice. As Christianity spread to these regions, the Church adapted this practice to the Advent season, reminding us that, just as the days are getting shorter and nights longer due to the Earth's orbit, so, too, does our world, on a spiritual level, experience the deepening of darkness through sin. We hear in the Gospel from Luke a powerful presentation of this twofold darkness.

> There will be signs in the sun, the moon, and the stars, and on earth nations will be in dismay, perplexed by the roaring of the sea and the waves. People will die of fright in anticipation of what is coming upon the world, for the powers of the heavens will be shaken. And then they will see the Son of Man coming in a cloud with power and great glory. But when these signs begin to happen, stand erect and raise your heads because your redemption is at hand. (Luke 21:25–28)

Although this Gospel may sound foreboding, not all is doom and gloom in the Advent season. One of the main themes of this season is that hope will return as our spiritual "nights" become less "dark" and our days become a bit "brighter." And what is the source of this hope? Our hope is the light that shines in the darkness: the person of Jesus Christ.

One way of understanding why we celebrate Jesus' birth around the time of the winter solstice is to embrace the Cosmic Liturgy image of creation. This vision, given its clearest voice in the writings of Maximus the Confessor and later developed in the works of Hans Urs Von Balthasar and Joseph Ratzinger, argues that every aspect of our world reflects a powerful metaphor of Christ's work of salvation.

The beginning of the shortening of days is accompanied by two events: the summer solstice and the feast of the birth of John the

Baptist. The birth of John the Baptist, placed on June 24, reminds us of John's call for the repentance of sin before the coming of the Messiah. The natural rhythm of the shortening of our days after the summer solstice is seen as a reminder of the effects of sin from which John calls us to turn away.

When Christmas finally arrives, our days begin to lengthen again, signifying that the light of Christ has come into the world through the Incarnation. This approach to understanding why Christmas is celebrated at the time that it is provides much more spiritual meaning than trying to sift through the debate about the actual date that Jesus was born, which, to be quite frank, wasn't all that important to early Christians.

Over time, this vision of the Cosmic Liturgy was lost, making many practices seem obsolete and, therefore, abandoned. For example, those of you who are more "seasoned in life" may remember the Tridentine Rite in which the priest stood with his back to the people while celebrating Mass at the altar. At the time of the Second Vatican Council, there was a concern that this and other practices were alienating the faithful from the priest during Mass, creating an air of clericalism. Yet, this does beg the question: Why did the Church do this in the first place?

The answer is rich and complex, but, essentially, the position of the priest was not meant to separate the clergy from laity. Rather, it had more to do with the understanding of the Cosmic Liturgy. In early Christianity, all churches were "oriented" toward the east, meaning that if you were sitting in the pews as a parishioner facing the altar, you would be looking east. The rising of the sun was seen in the early Church as a powerful symbol of the rising of the Son, Jesus Christ, from the tomb (hence, the importance of a "Sunrise Service" on Easter Sunday). In light of this, (no pun intended) the priest and people would face east to symbolically "orient" their prayers to the risen Christ, not seeing the sun as a god, but allowing it to be a powerful symbol of the risen Christ. This is but one of many examples of how the early Church sought to explore the intimate connection between God and creation.

In our modern times, we can see that the early Christians desired to express their faith in a way that connected every part of creation with God. The rhythm of the seasons became a way to not only mark time, but to provide hope amid life's darker moments. Nature itself, as brutal as it can be, gives us signs of new life, analogous to the new life we find in Jesus Christ. Despite the fact that our understanding of the mechanics of the universe have showed obvious errors in how the early Church viewed the world, the spiritual message of Advent endures: Come, Lord Jesus, and dispel the darkness of our lives with the light of your love, mercy, peace, and justice!

What is the lesson we can learn from this reflection? Living in a region that has well defined seasons, Advent reminds me that life has a rhythm of birth, growth, blooms, maturation, harvest, death, and new birth. The world we live in experiences great darkness, whether it be acts of terrorism, war, mindless shooting of the innocent, or discord within our homes and communities. These events can evoke in the human heart an underlying narrative of despair, thinking that all hope is lost in our world. Advent reminds us that, even in this darkness, there is hope.

At times, the seasons associated with the pilgrimage journey of our common home around the sun can be filled with darkness, cold, and despair. However, just as the Earth has a way of allowing all parts of its surface to be bathed in the light of the sun, so, too, does God allow all people, believer and non-believer, to experience the love of the risen Son in their lives.

This holds, for me, one of the greatest mysteries of the Advent season: In order for the love of the Son to be shown to the entire world, we must allow the "light of Christ" to shine in us.

Spiritual Exercise: Do you wish, in the words of Isaiah, to be a people who remain in darkness or are you willing to let your eyes see a "great light," awakening new life in you through Jesus Christ? Open yourselves to this light and, together, let us prepare our hearts to meet the risen Lord!

CHRISTMAS

HE IS BORN!

CHRISTMAS! IT CAN BE the most wonderful and stressful time of the year. While shopping for gifts, decorating our homes, or making plans to be with family, it's easy to forget the reason we give so much attention to this time of year: The celebration of the Incarnation.

> Now there were shepherds in that region living in the fields and keeping the night watch over their flock. The angel of the Lord appeared to them and the glory of the Lord shone around them, and they were struck with great fear. The angel said to them, "Do not be afraid; for behold, I proclaim to you good news of great joy that will be for all the people. For today in the city of David a savior has been born for you who is Messiah and Lord. And this will be a sign for you: you will

find an infant wrapped in swaddling clothes and lying in a manger." And suddenly there was a multitude of the heavenly host with the angel, praising God and saying: "Glory to God in the highest and on earth peace to those on whom his favor rests." (Luke 2:8–14)

Amid the Christmas parties, plays, concerts, and movies, there is an inner need to create a space of contemplation and silence to reflect, as did the shepherds, on this great mystery. Something that aids our contemplation is the crèche, with Jesus placed in a manger while Mary and Joseph reverence the newborn Messiah. In December for 2015, my astronomer's heart was pleased to have both a full moon and a rather peculiar, two-tailed comet named Catalina join the Holy Family. There are few things more satisfying to a hobby star-gazer priest than having wonders in the night sky accompany high points of celebration for the Church, especially at Christmas!

You may wonder, was there any Christian significance to a full moon on Christmas? Nope. How about some gloom and doom with a "Christmas Comet?" No way, no need to go there. Despite the irritation of seeing online conspiracy theories of end-of-the-world nonsense, I encouraged my parishioners to sit back and enjoy these nighttime wonders. For me, the only significance these astronomical events offered was personal: The memories of the last Christmas Eve full moon in 1996.

I was halfway through my "second senior" year at the University of Wisconsin-Stevens Point and was home for Christmas. I woke up in the middle of the night and knew I wasn't going to be able to get back to sleep for a while. Quietly, I walked through my parents' house, trying not to wake anyone. As I looked out the kitchen windows, I was struck with how bright it was outside, a result of the light from the full moon and its reflection off the snow. I slipped on my coat and boots and went outside to look at the moon.

The night was still and silent. I leaned on the railing of our deck and looked across the fields of our central Wisconsin farm. I was struck by how the moon made everything so bright that it almost felt like day. Suddenly, I heard the sound of something off in the distance briefly cry out, breaking the silence. It was probably a fox. The cry echoed through the night like the harmonious decay of a choir that arrives at the final cadence of a motet. Something grabbed my emotions in that moment and prompted me to ask, "Was that what it was like to hear the sound of the distant cries of Jesus the night he was born?" My heart rested in peace and I stayed there as long as my body would allow before the cold forced me back inside. I will never forget that mystical night!

Spiritual Exercise: What were the times in your life when God surprised you with a moment of His love and presence? Was it on a quiet, starry night or a warm summer day? Were you in the wilderness, on a city street, or in the quiet peace of a church? God has touched my life in these and other places. The science of astronomy has enriched my life and filled our world with wonderment, presenting us with fundamental questions about the world we live in. Astronomy also allows us moments to step back and simply gaze at the beauty of God's creation. Next Christmas, let's allow ourselves to be taken, once again, with the beauty and wonder of the universe and give thanks for the moment when Creator and creation came together in the mystery of the Incarnation. Let us listen in prayerful silence for the gentle cries of our Lord, Jesus Christ, calling us to embrace a life of faith, hope, love, and peace.

EPIPHANY

HOW ALL OF CREATION POINTS
TO OUR SOURCE AND SUMMIT

E PIPHANY RECALLS THE JOURNEY of three Magi who sought to find a newborn king, based on their understanding of the heavens. The drama is to realize they were not the only ones who were seeking Jesus. Herod, too, wanted to learn the whereabouts of this infant king, but for a very different reason: to eliminate a threat to his power.

In another narrative from the Gospel of Luke, the shepherds, representing the poor and the marginalized, are seeking out this gift from God after being serenaded by the heavenly hosts. Their motive, apart from the vision they received, is unclear. Perhaps, it was out of simple curiosity that they trusted what they had learned from their mystical experience. In the background of these narratives is the long-standing hope of the Children of Israel for a promised Messiah.

Others hoped for a military ruler who would eliminate an occupying force. Instead, Christ's battle cry was "love your enemies" and He found examples of faith in the oppressor.

When we take these narratives as a whole, what we discover is a dynamic tension in which everything, both carnal and incorporeal, was pointing humanity to a cave, a child, his mother, his foster father, and the manger. All people from the known world, Jewish and Gentile, rich and poor, kings and peasants, women and men, were looking, yearning to find the source of truth, goodness, and beauty. And when this source was found, it was wrapped in swaddling clothes, innocent, defenseless, born into poverty, and protected only by what must have been apparent to both Mary and Joseph, the grace of God.

Often, people ask the question at Christmas, "Was the star of Bethlehem real?" Others far brighter than I who have studied this matter more intently can handle that question. The question I find more interesting is this: What happened approximately 2,020 years ago that so changed the course of human history that even the way we measure the passing of time was altered?

Was the star real? Perhaps the more intriguing question to ask is whether or not "The Light" came into the "darkness," ending the perpetual Advent of those who yearned to see their hopes fulfilled? And when that light came, it brought about a new dawn, a "New Star" of hope in the lives of those who trusted in God, those who did not trust in God, those who studied the natural world, and those who were serenaded by the divine. Put simply, "Did God meet his people in the person of Jesus Christ?" This question has a little more teeth to it than simply going through old star charts and compiling theories about what was in the night sky at the time we think Jesus might have been born. This question asks *us*: Has our hope dawned?

Obviously, this question goes beyond the limits of science and points to a clear "leap of faith." However, to treat faith in such a pithy manner seems to be inadequate. For example, I have faith that Jesus Christ is my Savior, but I would not call that faith blind, unintelligent, unstudied, or divorced from natural reason (including the sciences). My faith has gone through times of profound light and darkness, ascent and doubt, consolation and desolation, but the end result has always lead me, no matter how "strong" or "weak" my belief has been, to profess one, noble truth: Jesus Christ is Lord – Who is the way, the truth, and the life!

To explore this further, let's reflect on the question posed by Pontius Pilate to Jesus: What is truth? The question of truth is something our world wrestles with to this day. For some, truth is simple. They resist questions that could cast doubt on what is presumed to be fact. For others, truth is illusive. They allow questions to multiply, deconstructing the thin fabric of childhood innocence that once wrapped them as a blanket but now exposes them to the "cold and chill" of doubt. Still others see truth as an adventure. They are willing to take off both the doubts and the protective "garments" of our innocence to seek clarity, exploring the most elemental questions of human experience. This journey can lead to complex philosophical abstractions, creating a fascinating, yet complicated web of reasonable arguments for truth. Some people give up the pursuit, finding the road frustrating and cumbersome, presuming there is no end to it. They become complacent, presuming truth cannot be achieved.

For the Christian, all of these reflections can be present, but there is one additional dimension unique to Christianity that transforms our lives: Truth is found in an infant who is far more than an infant. On the feast of the Epiphany, truth is found in Jesus Christ and, in this discovery, we receive an invitation to do far more than accept a chain of rational arguments to prove that something exists. We are invited to enter a new journey in which we find truth, goodness, beauty, faith, hope, and love through a relationship that transcends any type of human bond we can imagine. We are called to find a relationship that is like a well of water that never runs dry or bread from Heaven that always feeds our hunger. This water that never leaves us thirsty and this bread that always feeds us draws us into a bond of love whose permanence transcends this world, connecting us to the eternal love that can only come from the source of all things.

In short, one of the ways to approach the feast of the Epiphany is to see that every aspect of creation is pointing to the coming of the source and summit of all creation. Whether it be natural reason or Divine Revelation, all of the avenues of truth at the time of Jesus' birth pointed to the source of all truth. These two "wings" that allow our soul to ascend to God remind us that this journey is not limited to an elite class of people, but is accessible to everyone, according to his or her ability, to enter into a relationship with the source of truth.

So, whether you're a "Magi" who seeks for truth through natural reason, a "Shepherd" who is responding to a call from beyond, or a little bit of both, we all find ourselves on a journey, traveling different paths, leading to the same end: The God who is our Source and Summit.

Spiritual Exercise: How do you come to truth? Are you more of a "Magi," gravitating toward natural reason? Are you a "Shepherd" who is compelled by Divine Revelation? Or are you a little bit of both? Our faith calls us to understand that both faith and reason are necessary for the soul to ascend to God. And in that ascent, may we come to know the source of our journey when we celebrate the feast of the Incarnation.

THE TRANSFIGURATION

THE LIMITS OF LANGUAGE AND
THE POWER OF METAPHOR

How do you put into words that which goes beyond words? On the Second Sunday of Lent in 2015, this question was deeply entrenched in my heart. To begin with, I was deeply moved while trying to visualize the Biblical scenes of God, Abraham (Abram), and the invitation to Covenant that would bind them and future generations.

> The Lord God took Abram outside and said,
> "Look up at the sky and count the stars, if you can.
> Just so," he added, "shall your descendants be."
> Abram put his faith in the LORD,
> who credited it to him as an act of righteousness. (Genesis 15: 5–6)

As a child, I loved gazing into the night sky at our family farm. The combination of warm June nights and minimal light pollution allowed for wonderful views of the stars. On moonless nights, the "cloudy" appearance of the Milky Way was stunning. When I bought my first telescope in college, I would gently move the optical tube back and forth, making my way through the endless sea of stars. I often thought of Abram in those moments and wondered, "What was it like for Abram to look at the stars and realize God's love for him?" It's exciting to think that the advancements of astronomy have done nothing but affirm and deepen the core lesson of this metaphor from Genesis: Just as God has blessed the universe with an unimaginable number of stars, so, too, will God's Covenant with Abram and us be unthinkably fruitful! This returns me to my initial question, "How do you put into words that which goes beyond words?"

The Gospel speaks of another unspeakable event: The Transfiguration of Jesus Christ. This event is what theology calls a "Theophany," a visible manifestation of God's presence in the world. Now, being that Jesus is the Second Person of the Trinity, one could rightly argue that each moment of Jesus' earthly ministry was a Theophany. However, the event of the Transfiguration is unique amid the many events of Jesus' life.

Jesus took Peter, John, and James
and went up the mountain to pray.
While he was praying his face changed in appearance
and his clothing became dazzling white.
And behold, two men were conversing with him, Moses and Elijah,
who appeared in glory and spoke of his exodus
that he was going to accomplish in Jerusalem.
Peter and his companions had been overcome by sleep,
but becoming fully awake,
they saw his glory and the two men standing with him.
As they were about to part from him, Peter said to Jesus,
"Master, it is good that we are here;

let us make three tents,
one for you, one for Moses, and one for Elijah."
But he did not know what he was saying.
While he was still speaking,
a cloud came and cast a shadow over them,
and they became frightened when they entered the cloud.
Then from the cloud came a voice that said,
"This is my chosen Son; listen to him."
After the voice had spoken, Jesus was found alone.
They fell silent and did not at that time
tell anyone what they had seen. (Luke 9:28b–36)

This passage is rich in meaning and symbolism. To start with the basics, what does Scripture mean when it says that Jesus' "face changed in appearance" and his garments became "dazzling white?" These references are clearly trying to point to something that is beyond our comprehension, using human language and imagery to help the mind grasp that which cannot be grasped. Further, we see the figures of Moses and Elijah present with Christ in this "new way of being." Why are they there and what do they signify? Moses is the figure that signifies the Law of the Torah, and Elijah is the figure that signifies all of the Prophets. Therefore, between the Law and the Prophets stands the realization and fulfillment of both: The Person of Jesus Christ.

The pinnacle is reached when a voice is uttered amid a cloud that descends upon Peter, John, and James: "This is my chosen Son, listen to him." Was the reference to the cloud a change of barometric pressure or is it something else, a different kind of "cloud" that can hinder not only our physical sight, but also the eyes of the soul?

As God's voice affirms Jesus as the beloved Son, can we not hear in this cloud an affirmation of God's love for us, claiming us, through Baptism, as chosen daughters and sons? This "mountaintop" moment for Peter, John, and James was not only meant for them, but also for us, calling us to a "theophany of the heart," creating space for God to "speak" divine love within us in a way that goes beyond human words.

This passage also foreshadows one of the greatest mysteries of our faith: The Resurrection. How do you express the resurrection in human terms? What construct of human imagery can we devise to give us a glimpse of what this reality will be like? Will it be like Jesus' transfiguration, glorifying and transforming our existence into a new state in which time ceases and we enter an eternal state of being? Will it be like gazing naked-eyed into the night sky and being able to see the totality of creation in a way that would make the Hubble Space Telescope look like Galileo's first telescope? Will it be like the newly engaged couple who drops by my office, brimming with joy, simply wanting to share the love they have discovered, inviting me into that journey as the priest to celebrate their wedding? Is it like a young boy or girl who, while praying, is reminded of God's love for them through the words spoken of Jesus at his baptism, this is my chosen son, with whom I am well pleased? The answer to these questions is that the experience of the resurrection may hold a bit of all of these moments, but also contain joys so beyond anything we can experience that human words are inadequate.

Of the many blessings I have received from writing for *The Catholic Astronomer*, one of them has been deepening my awareness of how we can come to understand God through understanding creation. That being said, I also learned that there are limits to this understanding, pointing to a transcendent reality in which all language and human expression ultimately breaks down.

Spiritual Exercise: I would invite all of us to reflect on the question, "How do we put into words that which goes beyond words?" On a clear night, go out to a dark place where you can appreciate the night sky. Read Genesis 15, place yourself in Abraham's shoes, and then look up. Try to count the stars and see in this sign of the stars a constant reminder of the love God had for Abraham, the love God has for you, and the love God has for every person who has ever existed and who will exist. The next day, hike a hill you feel comfortable summiting and pray with the Transfiguration from Luke 9:26–36 and imagine what it would have been like to be present with Jesus, Moses, Elijah, and the Apostles. Together, may we wonder at the love God has for us.

EASTER VIGIL

THE BEAUTY OF THIS NIGHT

WHAT IS THE HOLIEST night of the year? Many would answer Christmas Eve. The correct answer, however, is Holy Saturday or the night before Easter Sunday. It is this night that Christians celebrate the resurrection of Christ from the dead. Not only is this night the basis for our faith that Jesus Christ is the Son of God, it is also the basis for our hope in the bodily resurrection. What I find beautiful about Christmas and Easter is that both celebrations sanctify that which many cultures cast as a negative: The night.

Now, night can symbolize sinfulness. As I have shared before, one of the reasons the Feast of the Birth of John the Baptist is celebrated on June 23 is that it is at about this time (in the northern hemisphere) that the days start to become shorter, providing a natural symbol to the Christian of a world that falls more and more into sin. December 25 roughly coincides with the time when the days get longer, giving us a natural symbol of hope that comes with the light of Christ. Christ's birth, now, sanctifies the night, taking that which was a symbol of sin and transforming it into a symbol of new hope like a star shining in the night sky. This leads to a logical question: What, then, is the symbolic interplay between light and darkness at Easter?

Let's look more closely at Holy Saturday, one of the liturgies of the Church year that must be celebrated after sundown. By celebrating at night, we honor the Jewish tradition of viewing the start of a new day at sundown instead of midnight. Another reason we do this is to allow the night, once again, to be a powerful symbol of the power of sin that so invaded the world that it put to death the Second Person of the Trinity.

The celebration begins with a simple flame, a light in the darkness, representing the light of Christ (Lumen Christi). After the blessing of the fire, there is a procession into a dark church that is, at first, illuminated only with the light from the Easter candle, but then is slowly filled with a warm light as each parishioner lights a candle from that Easter flame. The symbolism is hard to miss: The light of Christ illumines the darkness of a sinful world.

Similar to how the birth of Christ transforms the symbol of the night from despair to hope, so, too, does the light of Christ's resurrection transform the night before Easter Sunday from a symbol of defeat to ultimate victory. This transformation of symbol is present elsewhere in Christianity, such as the transformation of the symbol of water from its Old Testament meaning of sin and death to the waters of new life. This transformation of symbol is accomplished through Christ's baptism, signifying his entry into our human condition, transforming the symbol of water to one of new birth that, when combined with the Trinitarian formula of the baptismal rite, washes away our sins, incorporates us into Christ, infuses virtue into the soul, and gives the recipient the right to heaven.

During the Easter Vigil, one of the more powerful expressions of Christ's resurrection is the ancient hymn, the Exsultet (Praeconium Paschale). The hymn sings of how the heavens and the earth are to rejoice that Christ has risen. We hear the constant reference to "this night," reminding us of Salvation History, starting from the "happy fault," the "necessary sin of Adam," which paved the way for the coming of the one, true "Morning Star" of Jesus Christ to illumine our darkness. Adam's sin is a "happy fault?" Again, this poetic language draws out a great mystery of Christianity that life in Christ often takes what appears to be defeat and changes it to glory through Jesus' total gift of self.

I'm always moved by the section that speaks of the "things of heaven" that are "wed to those of earth," recalling a more ancient metaphysic of connecting the night sky with the heavenly realms. Although modern science has obviously shown this ancient worldview to be in error in the material sense, I can't help but think there is something we can recover spiritually from this image that reclaims a vision of all of creation as a sacred gift from God.

This odd meeting point of symbolic tension that is resolved through the light of Christ also reminds us of the inner battles we face as individuals and as a community. We live in a world that increasingly focuses on the symbols of sin and death, leading some to despair that either evil is prevailing or violence must be used to overcome the darkness. It is in these moments when our hearts can enter a metaphorical "tomb" with the body of Christ. The mystery of the Easter Vigil reminds us that the darkness of the grave is temporary and will eventually give way to new light through the resurrection of Christ. This is what inspired people at other dark times of history to not lose hope, but turn to the illuminating light of Jesus Christ, stirring the faithful to confront evil with love and peace. A prime example of this is the oft-quoted words of Martin Luther King Jr. from his work, "Where do we go from here? *Chaos or Community.*"

King writes, "The ultimate weakness of violence is that it is a descending spiral, begetting the very thing it seeks to destroy. Instead of diminishing evil, it multiplies. Through violence you may murder the liar, but you cannot murder the lie, nor establish the truth. Through violence you may murder the hater, but you do not murder hate. In fact, violence merely increases hate. So it goes. Returning violence for violence multiplies violence, adding deeper darkness to a night already devoid of stars. Darkness cannot drive out darkness; only light can do that. Hate cannot drive out hate: only love can do that. ("Where do we go from here? Chaos or Community." [1967] p. 62–63)

Spiritual Exercise: Let us bring the "dark nights" we encounter to the light of Christ's resurrection. Let us not view the night as merely a symbol of sin, but let us gaze into the night with wonderment, realizing that Christ has sanctified the night through his victory over sin and death. As a hobby astronomer, it's rather easy for me to see the night as sacred, imagining the stars and planets I can gaze upon with the naked eye as reminders of the hundreds of candles that burn at the Easter Vigil, taking their light from the Paschal flame. And as we move from a romantic introspection of the Easter Vigil to the difficulties of the real world, may we work to ensure that the sacred night Christ has inaugurated through His resurrection may never know a "complete darkness" in which the light of our faith ceases to blaze. May we work for peace, confronting the violence of this world with love, the love that is modeled for us through the example of Jesus Christ.

SACRAMENT OF CONFIRMATION

FEAR OF THE LORD – A GIFT OF THE HOLY SPIRIT MEETS ASTRONOMY

"THE FEAR OF THE LORD is the beginning of wisdom; prudent are all who practice it. His praise endures forever." Psalm 111:10

The months of April and May mean one thing for many Catholic churches in the United States: It's Confirmation Season! The crack of a bat on a baseball field might signal the beginning of spring, but the smell of sacred chrism being applied to the foreheads of Confirmandi is one of the final major events before parishes ease into the calmer schedule of summer. Confirmation can be a hectic time of coordinating spring sports schedules, finishing service projects, and making sure the Confirmandi show up on time for the big day. Amid all this Confirmation planning, we revisit a passage from Isaiah that reminds us of the gifts given at this celebration: the gifts of the Holy Spirit.

...a shoot shall sprout from the stump of Jesse, and from his roots
a bud shall blossom.
The spirit of the LORD shall rest upon him:
a spirit of wisdom and of understanding,
A spirit of counsel and of strength,
a spirit of knowledge and of fear of the LORD,
and his delight shall be the fear of the LORD.
Not by appearance shall he judge,
nor by hearsay shall he decide,
But he shall judge the poor with justice,
and decide fairly for the land's afflicted. (Isaiah 11:1–4)

As beautiful as this passage is, I always struggled with the idea of the fear of the Lord as a gift of the Holy Spirit. Growing up in a family that strongly affirmed a loving God that is merciful and forgiving, the idea of fearing God seemed out of place. However, as the psalm at the beginning of this reflection states, not only is the fear of the Lord a gift of the Holy Spirit, it is the beginning of wisdom and holiness. The beginning of holiness means to be afraid? I was in need of some clarification.

The search for clarification began a journey that has borne much fruit. The first stage of this journey came in seminary when I encountered the idea of fear of the Lord being connected with liturgical reverence, expressed in awe and wonder. At first, I was comforted by the image that God was not something I needed to be afraid of, but rather someone I reverence and honor. Over time, however, this understanding of fear of the Lord began to leave me dry. Perhaps it was because I sang in the seminary choir and spent many a liturgy with a folder in my face, but my participation in Mass seldom put me in a position where I could step back and be taken in by awe, wonder, and reverence. In fact, the only time I felt I could appreciate quality liturgy was after Mass was done.

The next phase of this exploration came as a priest/teacher at Regis

Middle and High School in Eau Claire, Wisconsin. In the fruitful struggle of teaching abstract, theological concepts to students who had yet to develop the ability for abstract thinking, I found a concrete way of teaching the fear of the Lord through the 10 Commandments.

With my freshmen, I would argue that the 10 Commandments should not be seen as weighty rules that oppress our fun. Rather, the Commandments are boundaries we place on our relationship with God, similar to how we protect meaningful friendships out of fear of hurting a person we love. That means the fear of the Lord is not something that should make us shrink in terror, but became a positive process of protecting the most essential relationship of our lives: our relationship with God.

I further applied this "spiritual personalism" (for lack of a better term) when trying to help my students understand Hell. I would explain that Hell was not a place of molten lava, pitchforks, and heavy metal music, but rather a state of being, best understood as a radical, eternal loneliness and separation from anything that is true, good, and beautiful. Although it was impactful to recast the fear of the Lord into something that motivates one to protect something versus dreading something, this relational understanding still didn't address the aspect of awe and wonder we are to have toward God for simply being God. So, I kept searching.

Recently, my pondering of fear of the Lord has been aided by astronomy. When I gaze on objects like the Orion Nebula, contemplating how it is a place of star birth, made possible through star death, and reflect on the beauty of its rich color, the experience evokes the sense of the reverential awe and wonder I was taught to embrace in Sacramental Theology.

It's ironic that a literal distance is necessary to experience this awe and wonder, along with the patience and time to gather the necessary photons through a camera to display the true wonder of these objects (with a little artificial enhancement from iPhoto). Just as the spiritual personalism of the 10 Commandments helped me to experience God's imminence by protecting a crucial relationship, so, too, does viewing God's creation through a telescope give me a sense of God's transcendence, helping me to fix my spiritual gaze on the love of the Trinity and moving me to awe and wonder while contemplating God's beauty and majesty.

As a child, I would often wonder, "What would a star look like up close?" When I got older, I learned how to safely use a solar filter to gaze at our nearest star, the sun. (I emphasize the proper use of solar filters. Do not look at the sun without one of these filters and the help of someone who can teach you how to us it properly!) Every time I gaze into the sun, I experience great awe and wonder. However, common sense tells us we need a safe, "reverential" distance to have this experience.

This reverential distance reminds me of the radical transcendence in which God is profoundly other, our source, our beginning, and our end — concepts too big to fully grasp with the limits of human reason. Yet, this transcendence can impact us on a personal level through gazing at solar events, like a prominence exploding off the sun.

Transcendence and imminence, two dynamics of the universe that remind us of ways God is present to us. These connections, however, need to avoid a crude pantheism, turning the metaphor into a "nature god." Instead, we gaze in wonder at natural transcendence and imminence, realizing these same categories pertaining to God go far beyond anything we can comprehend in the natural world.

Exploring the fear of the Lord through imminence and transcendence has reawakened my sense of awe and wonder during Mass and in my daily life. Further, it has helped me appreciate God's fingerprints present in the beauty of creation and calls me to constantly work toward upholding the human dignity of my neighbor. In short, I finally feel that I am approaching a healthy understanding of the fear of the Lord as the *beginning* of wisdom.

Spiritual Exercise: How do you understand the gift of the Holy Spirit we call "the fear of the Lord?" Do you feel a sense of awe and wonder in the presence of God? Do you feel a sense of awe and wonder when you gaze into the heavens? My prayer for you is that you will find a language to express the beginning of wisdom that helps deepen your love of God.

MEASURED TIME AND SACRED TIME

SO, WHEN IS EASTER?

How can you reduce your priest to a stammering ball of confusion with one, simple question? Ask, "Father, what is the date for Easter next year?" Unless your pastor had the good fortune of looking at his liturgical planner a year ahead of time, I can almost guarantee his answer will be, "Umm... I think it's... ah... late March... early... mid... or late April?" The reason this question can stump the smartest of liturgical chumps is that the date of Easter is a collision point between differing approaches of measured time.

The calculation of Easter is done by identifying the first Sunday that falls after the first full moon after the vernal equinox. Sounds easy enough. However, given that lunar cycles do not match up exactly with our calendars, we can have a range of dates for Easter. For Western Christians, those dates can occur between March 22 and April 25. For Eastern Christians, the dates fall anywhere from April

3 to May 10 because they follow the Julian calendar instead of the reformed Gregorian calendar. Therefore, the best way to calculate the date of Easter is to write the Vatican Observatory and request its yearly calendar!

This collision of differing measurements of time begs a question: How do we understand time? When I was at the first Faith and Astronomy Workshop in January of 2014, Fr. Paul Gabor gave a fascinating presentation on "leap seconds." We learned that no two days are ever the same because of the irregularity of the Earth's rotation. Therefore, the correction of "leap seconds" must be made to ensure that your GPS gets you to your destination and that groups, like the military, that use GPS in targeting systems can hit what they intend to hit. Because one second of earth's rotation represents about a kilometer of distance from one point to another, one can see that being off by a second can even be a matter of life or death.

The measurement of time was not always so mechanical. Fr. Gabor explained that the ancients measured time primarily through significant events that merited religious and cultural celebration we can call "sacred time." In modern society, our measurement of time has slowly divorced itself from observing these moments of significance. Holy Week, unfortunately, is becoming a prime example of this divorce. Gone, for many, are the days when schools and businesses would close to allow their employees and students the time to observe Good Friday. Instead, the Paschal Mystery must be "fit in" around work schedules and school projects, due to the cultural amnesia toward the greatest event in human history: The Resurrection of the Son of God.

In light of this, we can ask another question: How should we understand time? One may argue that we should only focus on "sacred time," emphasizing the moments of profound religious and cultural meaning. Others say that religious calendars are arcane relics of a dying world view and that the mechanical passing of time, divorced from religious significance, is all that is needed. Which of these mentalities is correct? I propose that the Catholic "both/and" principle can help us find a balanced solution.

I find it ironic that both measurements, sacred and mechanical, are imperfect and need adjustments to properly measure time. There is no such thing as a "perfect" day. Both measurements of time have their place in society. When we lose the sense of the sacred that comes with feast days, we forget a fundamental part of who we are as a people and can drift from joy to cynicism. These significant moments are sacred in part because of the passing of time that distinguishes the sacred from the mundane (or profane), ordinary living of everyday life. In short, the interplay between sacred time and mechanical time helps us appreciate the reason we measure time in the first place: to reflect on life's meaning and purpose.

Spiritual Exercise: How do you find meaning and purpose in a "leap second" world? What are the struggles you face to embrace the "two calendars" of sacred time and mechanical time? Reflect on these questions and may the Lord help us to use our time wisely to build up the Kingdom of God.

BEGINNINGS AND ENDS

O DEATH, WHERE IS YOUR STING?

"A MEN, AMEN, I SAY to you, unless a grain of wheat falls to the ground and dies, it remains just a grain of wheat; but if it dies, it produces much fruit." (John 12:24)

Do you wish to advance in the spiritual life? If so, you must "die to self." This basic axiom of spirituality seems counterintuitive: In order to find new life we must experience a type of death? Those who have committed themselves to the spiritual life will quickly affirm that new life comes through death. Whether it be confronting a moral struggle, overcoming our fear of doing God's will, or embracing a life of prayer, there is first an inner moment of humbling one's self to become the "grain of wheat" that falls to the ground and dies before the "tender green shoots" of new life in Christ begin to emerge. When approaching this spiritual death, one often experiences fear and resistance toward the unknown of who they will become at the end of this journey. However, after completing the journey, one discovers great beauty in the process of dying to self that was necessary for new life to take root.

As Christians, this dying and rising should be no surprise in light of the life, death, and resurrection of Jesus Christ. One of the prophecies most often quoted by Jesus himself was that he would be denied, handed over, suffer, die, and then rise on the third day. This mystery was met with resistance, not only by Christ's disciples, but by Jesus himself, expressed in a moment of hesitation when offering his prayerful distress to the heavenly Father with the words, "My Father, if it is possible, let this cup pass from me; yet, not as I will, but as you will." (Matthew 26:39b) Of course, we know the rest of the story, a story about the most stunning event in human history: The Resurrection. With the mystery of the empty tomb, we see the death of God in the person of Jesus Christ leading to a glorified reality, expressed beautifully and simply in the Book of Revelation, "Behold, I make all things new." (Revelation 21:5b)

Therefore, our spiritual journey of dying to self and rising to new life is a type of analogy of the mystery of the dying and rising of Jesus Christ. As deep a mystery as it is to experience a spiritual death that brings joy, peace, and healing, so, too, do we believe that our physical death is not the end and will lead, God willing, to eternal glory. Physical death is a part of life in which we pass through a type of "womb" between this life and the next. One can interpret each spiritual death as a type of preparation to help us face our physical death with the hope that where Christ has gone, we shall follow. Just as there is fear and anxiety with the spiritual dying and rising of the soul, so, too, there is great anxiety facing the mystery of what this "birth to eternal life" will be like.

As a priest, I have assisted many on this journey. Each death is unique and contains a whole range of emotions, fears, hopes, and joys for the person confronting death and the family that prepares to "let go" of a loved one. As illogical as this may sound, many times death can be given the title of "beautiful" when embracing the inevitable is not met with fear and trepidation, but with acceptance and courage to face the mother of all of life's journeys.

There are some whose acceptance and courage run so deep that they not only become comfortable with death, but in a real way befriend death, giving voice to one of the most profound passages of Sacred Scripture: "Where, O death, is your victory? Where, O death, is your sting?" (1 Corinthians 15:55)

These reflections often come to mind when I gaze at the beauty of an emission nebula. The soft pinks of a gas cloud tell a story of death, yet they are displayed in stunning beauty. Just as the fall leaves paint the countryside with beautiful reds and yellows before the onset of winter, so, too, do these remnants of stars paint the night sky as if it were the canvas of a contemporary artist.

In this story of death is also the story of new life in the birth of stars, shining like small jewels in the night. The mystery deepens when we realize that without the death of these stars, the elements needed for life to exist on our planet would not be present. There is a necessity to this death, similar to the necessity of a troubled soul needing to die to self, in order for new life to exist. Is it any surprise that the great mysteries of life, death, and what comes after death should be made present to us in the natural world? Does it not seem fitting that these grand images of nebulas are presented to us like icons hanging on a wall, drawing us into the great mystery and hope that our death, too, may be beautiful and give way to new life with God in the state of being we call Heaven? And in these luminous clouds of gas do we not see a metaphor for the life, death, and resurrection of Jesus Christ?

These sobering reflections give voice to the sentiment expressed in the third section Psalm 103.

For as the heavens tower over the earth,
so his mercy towers over those who fear him.
As far as the east is from the west,

so far has he removed our sins from us.
As a father has compassion on his children,
so the LORD has compassion on those who fear him.
For he knows how we are formed,
remembers that we are dust.
As for man, his days are like the grass;
he blossoms like a flower in the field.
A wind sweeps over it and it is gone;
its place knows it no more.
But the LORD's mercy is from age to age,
toward those who fear him.
His salvation is for the children's children
of those who keep his covenant,
and remember to carry out his precepts. (Psalm 103:11–18)

Spiritual Exercise: Pray with the mystery of Christ's life, death, and resurrection. Do you find hope in the Jesus' Paschal Mystery? My hope is that all of us can discover new life in Christ so that we may savor the eternal bliss that is promised in the life to come for those who love God.

CONCLUSION

R EFLECT ON THE QUESTION: How can the view of Cosmic Liturgy help enrich your personal prayer life and the prayer life of the Church? From my perspective as a Catholic priest, I appreciate the Cosmic Liturgy because it calls all of us to see every moment of our lives as a potential encounter with God. In this encounter, we begin to discover our dignity as Children of God, exploring the fundamental questions of life and death. In our final section, we will explore some of the important scientific events from the years 2015 and 2016 and offer some spiritual meditations meant strictly for your personal enjoyment and enrichment.

INTRODUCTION

SECTION FOUR

I N THIS SECTION, I present a mix of significant astronomical events for the years 2015 and 2016 and reflections on Scripture, science, beauty, and astronomy. The central event of 2015 was the historic flyby of the New Horizons mission to Pluto, giving us, for the first time, a clear view of this fascinating Dwarf Planet. Whether it be this achievement, the discovery of gravitational waves, or the 2014 European Space Agency's Rosetta Mission, which landed the Philae Lander on a comet, a common, practical question can be asked in the face of these monumental achievements: Does astronomy matter and why are we spending so much money on it?

Section Four begins by exploring this question from the standpoint of how "practical" astronomical research is and what benefit it provides for humanity both scientifically and spiritually. From here, we will explore philosophical questions about life on other planets, the role of beauty in our understanding of creation, the discovery of gravitational waves, the Pluto flyby, the transit of Mercury, and a series of spiritual reflection meant to foster a sense of awe and wonder.

FAITH AND ASTRONOMY

ARE A WASTE OF TIME!

OR... MAYBE NOT

FAITH AND ASTRONOMY ARE a waste of time! Although I obviously disagree with this sentiment, there are some who claim that faith and astronomy are not significant because both lack the ability to make concrete, functional advancements to society. Critics acknowledge that faith and astronomy may be great at exploring things like wonder and awe but they question what these fields contribute to society on a practical level. When these arguments are made, it's a painful reminder that we have become quite utilitarian in our worldview, valuing things more if they have a functional or usable purpose, while devaluing things that are more meaning driven and explore realities that are not easily measured in earthly terms.

This mentality not only influences faith and astronomy, it impacts many parts of the human experience. For example, when my grandmother, Edna Riley, developed medical issues that forced her to give up her driver's license at the age of 97, she experienced two deep emotional struggles. The first struggle was losing the freedom to travel. The second was deeper, expressed in the heart-breaking words, "Jamie, there's no use for me anymore... I'm useless."

Deep frustration grew in my heart listening to my grandmother's words. This is the woman who essentially raised me for the first two years of my life when my mother experienced serious medical issues after my birth. This is the woman who, as a convert to Catholicism, taught me that it was never bad to ask questions about my faith, even if it took me down challenging roads. This is the woman who began teaching in a one-room schoolhouse in the year 1925 and continually received visits from past students who simply wanted to thank her after she retired. And this is the woman who, when I would invite college friends to our farm, would end up being a teacher again as we would sit around her lift chair like children, listening to mesmerizing stories about life during the Great Depression. Useless, just because she was 97? Far, far from it!

This and other experiences I have had with a utilitarian worldview remind me of Pope Francis constantly warning us not to become a "throw-away society." This warning is often cast as part of the ecological vision of Laudato Si', calling all of us to avoid wastefulness, protecting the gift of creation God has given to us to ensure access to natural resources for generations to come. However, I also find this notion of not being a throw-away culture intimately tied to his concern for the elderly and the youth.

When people asked Pope Francis at the beginning of his papacy, "What are the greatest challenges facing the Church today," his response was joblessness of the youth and the loneliness of the elderly. Many scoffed at this, wondering why he didn't say something about terrorism, war, or abortion. Over time, Pope Francis has addressed these and other pressing issues of the Church. However, I fear that the narrow lens often used to analyze Pope Francis has led many to forget that avoiding the temptation of being a "throw-away culture" is tied to all of the pressing social issues mentioned earlier and more.

Avoiding a throw-away culture not only speaks to ecology, it also reaches the vast expanse of our understanding of human dignity, reflected in the following ethical and moral questions.

> Are we throwing away our world, our people, our young, our old, refusing to see God's presence in our neighbor, refusing to see God's handiwork in creation, and gutting all sense of beauty and wonder from them for the sake of usefulness and utility? Do we view life as simply mechanistic, waiting or the "machine" to break down? Do we view the elderly as indispensable sources of wisdom and history for our society or are they simply a financial drain on our health care system that could be better served elsewhere? Do we view an unplanned pregnancy as a gift from God, despite the circumstances of the conception, that is deserving of love and dignity or is it a "curse" that limits freedom, contributing to the overpopulation of the world, likely to live in poverty, and therefore "optional" at best to the world?

There are many other questions we could explore, but the core sentiment of the ethical questions raised are the same: Do we view the world with dignity or utility?

Now, how does this inform us when looking at the role of faith and astronomy in society? First, we need to avoid the trap of implying that there is no usefulness in faith and astronomy. We can find profound examples of how each has contributed to a useful society.

Despite the objections that astronomy is a waste of money, time, and resources that we could apply elsewhere, we need to remember what astronomy has already given to our world. Although some may argue the space race was spurred by Cold War politics, wasn't it the dreams and wondering of scientists and people of good will that not only put humanity on the moon, but in the process opened the door to a whole new way of viewing our world and our place in the universe? Don't we remember that the satellites that make our creature comforts like GPS, cell phones, and DishTV possible would have never existed if we hadn't strived to reach for the stars, creating new technologies in the process that have been seamlessly integrated into our daily life? And isn't it true that if we were to literally "pull the plug" on astronomy that we may be denying ourselves the next generation of technologies and understanding that could improve human dignity in the world we live in?

What astronomy teaches us is that taking the time to "waste some time" to dream and wonder has led to innovations that have contributed to society in practical ways. However, astronomy does something else that goes beyond creating new widgets. It feeds the soul by exploring some of the most basic questions of life: Who am I, Why am I here, and How did all of this get here? Questions that science, alone, cannot answer.

Regarding faith, we can see a similar interaction between wonderment and function. When I was in seminary, I was introduced to the ideas of time we call Kairos and Chronos. Chronos is the day-to-day, hour-to-hour, minute-to-minute understanding of the passing of time. Kairos, on the other hand, is a sense of time that focuses on profound moments in which we enter into a unique relationship with time, being made present again to the events of Salvation History. The centerpiece of this understanding of Kairos occurs in the Liturgy, making Mass the ultimate "waste of time" both in the practical sense and in the theological sense. Liturgy teaches us that there are times in life that all we need to do is be in God's presence with no tangible goals of productivity in mind. Still, we also need to ask, "What has been the fruit of this entering into the timelessness of God's Presence?"

Can we not see in this wonderment of the love of God the creation of some of the most powerful moments of human history like the Christian roots of the Civil Rights Movement, which found its strongest expression in the words of Dr. Martin Luther King Jr.'s "I Have a Dream" speech? Do we not hear the echo of the hymn, "Amazing Grace," reminding us of the slave trade, a slave ship captain who viewed Africans as mere utility to be bought and sold, and how the captain of this slave ship came to turn away from his sins by contemplating the love and forgiveness Jesus had not only for him, but for the slaves that were on his ship? These "Biblical nights of wonder" lead us to the same conclusion: In order to act rightly in the world, we need to wonder in God's presence about our world.

In short, faith can take the core questions we identified in astronomy, "Who am I, Why am I here, and How did everything get here," and expose them to the wondrous love of Jesus Christ. This allows us to not only contemplate questions like "Who am I in relation to God," but also the question, "How am I to treat others and myself because of my love of God and God's love of me?"

These reflections remind me of Mother Teresa of Calcutta and her persistence in having her Sisters start each day before the Blessed Sacrament. This daily holy hour was to remind the Sisters that just as they gazed upon the presence of Christ in the Eucharist, so, too, were they to see in the poor and marginalized of this world "living tabernacles," having within them the presence of Christ, that was to be loved, reverenced, and nurtured, not because of any utilitarian gain they possessed, but simply because they were children of God.

I wish to emphasize that we must avoid another trap when exploring the "usefulness" of faith and astronomy: The trap of reducing faith to ethics and astronomy to applied science, meant simply to make more technological toys. Faith and astronomy, in their pure sense, initially ask non-utilitarian questions that point the mind and heart skyward, wondering "What is out there?" As this exploration unfolds, we find a tension between wonder and function, purpose and purposefulness, dignity and utility. In this tension, let us remember that we share in this struggle, wanting to contribute something tangible to our world, but we also desire to find an inner dignity that is detached from any accomplishments. This dignity is rooted in simply being a child of God, made in God's image and likeness. Put another way, astronomy inspires us to look "up," while faith inspires us to take that heavenward gaze and look "in" at ourselves.

From the day my grandmother shared her heartfelt struggle with me, I prayed that she would stop feeling useless and realize that she was an indispensable source of love and knowledge for our family, her former students, and our parish. I know, in the mystery of God's love, that my prayer was answered, perhaps not as "perfectly" as I desired, but in the way that was best for Edna. We must not reduce our world and people to usefulness. Let us help people find dignity simply for being loved into existence by God. And, in the realms of faith and astronomy, let us remember that we need to wonder about our world and our God to help us make a better world for everyone regardless of age, gender, race, country of origin, or state of life.

FULL MOONS

LOOKING FOR LIFE IN ALL THE RIGHT PLACES

I
S LIFE UNIQUE TO Earth? Whenever I am asked this question, I hesitate. The hesitation isn't because I don't know how I want to answer the question. Quite the contrary, I know exactly how I want to answer. The hesitation comes because I know that what people ask and what they mean to ask are often two, fundamentally different questions. The question, "Is life unique to Earth," in the scientific sense implies an exploration for any type of organic life that is not found on our planet, whether it be microbial life or something as complex as animal life. The question, "Is life unique to Earth," in the theological sense can mean what is implied in the scientific sense, but often hides a deeper question: Will we find life made in God's image and likeness outside of our Earth? At one level, with what we currently know of the universe, the honest answers to both approaches to the question is, "I don't know." However, it is becoming more and more

evident that the discovery of life that would be on par with simple microbial life could be possible within our own solar system. But there is a third question that can be embraced with a wholehearted "yes" from both science and theology. That question is this: Is the universe "alive?"

To delve into this question, I would like to dial my personal clock back to the 1996–1997 school year and the course Astronomy 205 – The Solar System. It was a wonderful exploration of our galactic backyard by Dr. Randy Olson from the University of Wisconsin-Stevens Point. As we explored planets, moons, comets, asteroids, and our sun, I was most fascinated by the moons of our solar system. To use a musical analogy, the planets seemed like a classical symphony orchestra: well-proportioned with logical orbits (for the most part), and appeared, even in their differences, to fall into nice, identifiable classes when looking at the inner solar system (small, rocky objects) in contrast to the outer solar system (large, gas giants).

If the planets are a symphony orchestra, then moons are like quirky, odd little Seattle grunge bands. Each moon seems so radically unique, revealing great differences from their celestial "cousins" in our solar system. I was first drawn into the quirky world of moons when studying Jupiter's moon Io. This tortured little ball of sulfur is constantly turning itself inside-out due to the gravitational forces exerted on it by Jupiter and the neighboring moons. This volcanic moon would make a good, but terrifying, metaphor for Scripture's references to Gehenna, where "their worm does not die, and the fire is not quenched." (Mark 9:48) OK, perhaps that was a little over dramatic, but the point of emphasis I would like to make is that studying Io showed me that bodies of our solar system, apart from the Earth, were not boring hunks of junk just floating around in space. They were dynamic, volatile, fascinating, and, in a real way, "alive" with activity.

From a perpetually erupting sulfur volcano to what appears to be a massive snowball, let's look at one of Saturn's moons, Enceladus. When I taught high school astronomy, I would joke with my students that this moon looks like a skier's dream with its white, snowy appearance and a gravitational force that is 1.1% of Earth's. It takes "catching some serious air" on a ski jump to a whole new level (although I realize skiing on Enceladus would not be possible for reasons that go far beyond a lack of snow). Enceladus' surface is ice. What is fascinating about this icy surface is that it's full of cracks, implying that some type of internal activity is creating these cracks. Furthermore, areas of mist shoot out through some of these cracks, revealing a liquid layer of its interior that is being forced out by something (most likely a massive water level that is heated by the gravitational forces of Saturn and the moon Dione). If there is water, heated water under the surface of icy Enceladus, could there also be simple forms of biological life, similar to what we find in the deepest recesses of our own ocean? This is a question NASA is exploring. To try to learn what may be beneath the moon's icy crust, NASA flew the Cassini probe through some of the escaping mist on Enceladus.

I chose these two moons for a specific reason. On Io, we see a world that is "alive" and quite active, but not able to support organic life as we know it on Earth. To say that Io is "dead" is a clear misnomer. It is quite alive with its volcanism and presents to us a fascinating moon to study and, through that study, can help us better understand our own planet. Enceladus, on the other hand, is also alive with geological activity and water spouts, and it also teases out the real possibility of simple life forms that may exist under its icy crust or perhaps even microbial life that is jettisoned out amid the moon's escaping mist. This potential, twofold reality of a moon that is not only alive, but may also contain simple life begs the scientific sense of the question, "Is there life outside of the Earth?" No one is even hinting at the possibility of anything approaching the uniqueness of human life in this exploration of Enceladus, but the possibility of basic life is real and is creating a lot of excitement for many in the science community.

One may ask, "If we do find simple life on other planets (or moons), how would this affect the Christian view of creation?" Although I do not consider myself to be an expert on this theological exploration, my initial reaction would be that the only impact it would have would be a positive one. Scripture already affirms that all things that have been made give glory to God by their very existence, a truth that can even embrace the sulfur volcanoes of Io. Would Christianity be shaken to its core if simple life were discovered elsewhere? My answer would be a confident, No, it wouldn't shake Christianity. If anything, it would enrich our understanding of what it means for God to be Creator. The "hymn of creation" from the Prophet Daniel is again applicable because the discovery of simple life would add one more voice to this course.

> Sun and moon, bless the Lord;
> praise and exalt him above all forever.
> Stars of heaven, bless the Lord;
> *praise and exalt him above all forever. (Daniel 3:62–63)*

In this brief excerpt from this much longer hymn from Daniel (3:50–90), I hope it's easy to see that simple life found outside of our planet does not possess any issues theologically for our understanding of God and creation. Now, let's look at the question of whether or not we will find life that is made is God's image and likeness outside of our planet. This question is a bit dicer and needs to be handled with greater care.

First of all, would the discovery of intelligent life on other planets bring into question the Biblical understanding of the uniqueness of the human person? The difficulties in answering this question go far beyond the biological aspect of life and into questions of essence, being, soul, and what it means to be made in God's image and likeness. For example, let's say, hypothetically, that we did find life that has the potential of being understood as consistent with human life on our Earth. How would this discovery impact our understanding of Original Sin? How would it impact our view of monogenesis of the human person and ensoulment? Would we be able to identify self-reflective thought that went beyond animal instinct and includes a sense of morality and ethics? Would this life have its own sense of religion, and how would this relate to the understanding of religion on our planet?

As you can see, these kinds of questions are not able to be answered by discovering chemical markers that hint of biological life upon a distant earth in the "habitable zone" of another sun. These kinds of questions can only be answered through a "close encounter" with another life.

I intentionally use this image of encounter with another life to tease out one last question for this reflection, Do we understand what life made in God's image and likeness means on this planet? The atheist reading this chapter may lodge an understandable protest, "Most of your questions about exploring human life have little to do with science and more to do with religion and philosophy!" I would agree, in part, with that assessment. The theological exploration of how we define life does have aspects that include science, but it also deals with that which goes beyond science. And herein lies the problem: How much of our understanding of the human person is "purely scientific" and how much of it is philosophical and/or theological?

Let's face it, the concept of the human person was around long before the modern sciences were even dreamt of in the human mind. Yet, the sciences have contributed to a deeper understanding of the human person. When trying to define, "What makes a human a human," do we focus on a biological understanding, a spiritual understanding, or do we affirm a third way, the way of Thomas Aquinas, in which we are "body/soul," meaning a unique, inseparable union of matter and spirit?

Should science, philosophy, or theology be removed from developing the definition of the human person? Absolutely not! Although differences may exist among people about the role of religion in our world, it would be intellectually dishonest not to include the rich textures religion has contributed to our understanding of the human person, just as it would be a grave mistake to reject what science has contributed to our understanding of the human person. Therefore, before we even contemplate an encounter with another type of life that may possess the distinction of being made in God's image and likeness, let us first understand human life and dignity on this "little rock," inviting faith and science into a fruitful dialogue that not only seeks to define the human person, but also embraces the rich exploration of what it means to "be human."

In this pursuit, we need to prepare for the future by taking a sober look at our broken past, understanding how we have failed to uphold human dignity in our world. This is necessary so as not to repeat previous errors, ensuring that a future encounter with another potential life, including encounters between lives on this planet, are marked with respect and dignity, not violence and hatred.

Do I think we will encounter life made in God's image and likeness in other areas of the universe? To be honest, I don't know. I do feel confident that, barring something miraculous, an encounter of this nature isn't going to happen any time soon. Even if biological markers were found on other planets through astronomical observations, even if powerful telescopes could "spy" on organic life on other planets, would this discovery be able to quickly ascend to the theological density of life made in God's image and likeness? Without a real "human" encounter with that which is discovered, no, it wouldn't be able to be verified.

Nevertheless, let's also avoid the trap of trying to place limits upon God's creative act. After all, we do believe and hope that, God willing, we will experience life apart from this earth in the Kingdom of Heaven, encountering those who have gone before us in love and have joined the Communion of Saints and the heavenly hosts. Although this is not a reality that can be measured by science, we affirm its existence. If we can affirm this state of being with God (while also affirming the "other" state of being in the afterlife), let us not fear the mysteries we will discover beyond our common home in the material world.

Whether it be something hidden beneath a great ice sheet on a moon or something that is in a far off galaxy that we can't even comprehend at this point, let us trust that the God who loved us into existence gave us the desire to explore such things because He desires us to encounter such things, helping us to better understand ourselves and deepen our love of Him. And in that exploration, may we always remember that before we try to find life made in God's image and likeness outside of our common home in a distant "galactic neighborhood," let us first embrace what it means to be truly human, in the best sense of what this means, in our own neighborhoods.

FIBONACCI NUMBERS

WHAT DO THEY TELL US ABOUT OUR WORLD (IF ANYTHING)?

1, 1, 2, 3, 5, 8, 13, 21, 34, 55…Yes, this chapter is about Fibonacci numbers.

The first time I heard about Fibonacci numbers, it blew me away! The sequence is a simple pattern of arithmetic starting with 1 + 1 = 2. After that, add the answer and the second number of the equation, 1 + 2 = 3. Then, continue the sequence over and over again (2 + 3 = 5; 3 + 5 = 8; and so forth). What you end up with is an infinite sequence of numbers that, amazingly, is found throughout creation from the shape of ancient fossilized sea creatures to spiral galaxies. After my initial fascination with this sequence, I became hesitant about the "meaning" of Fibonacci numbers when I discovered rather bizarre applications ranging from hyper-literalist "theo-science" to New Age "guru-ism." Between the extremes of my personal fascination and "click bait" craziness, Fibonacci numbers in nature still grab my curiosity to this day.

My curiosity with Fibonacci numbers began in college. In my undergraduate work, I studied music. I primarily focused on saxophone and voice, but toward the end of my college career I began to work on music composition. Dr. Charles Young, professor of music at the University of Wisconsin-Stevens Point, introduced me to the art of composition by analyzing Bach minuets, canons, and other works. In addition to being wrapped in the simple beauty of the pieces, I studied the constant ratios of notes and measures used: 1 to 2, 1 to 3, 2 to 3, 3 to 5, and so forth. I began to comprehend music in a new way and tried to apply this "musical math" to my compositions Growing up in a pop music culture fueled by emotion, it never dawned on me that my favorite songs were embedded by fascinating mathematical proportions like Fibonacci numbers.

Over time, I was drawn into the fascinating, but slightly maddening world of how we find mathematical ratios and proportions all over music, art, architecture, science, and nature. I read the book "Godel, Escher, and Bach: An Eternal Golden Braid" and became consumed with the idea of strange loops, composing my own two-, three-, and four-part canons. To put it simply, I was hooked on music composition and the math behind the music. (I still write a little on the side for the schools and parishes I've served.) Yet, the more I tried to be a "musical mathematician," the more I realized that even though I might have gotten the math of the music right, I have never written anything that approaches the simple elegance of a Bach minuet. This revelation made it clear that a key ingredient was missing not only in my music compositions, but in my understanding of what sequences like Fibonacci numbers mean on a practical level.

So, what is the missing ingredient when understanding the connection between the math of music and my compositions? I would argue, it's the existence of beauty.

I've encountered a number of musical works that display mathematical brilliance. Whether it be classical or contemporary, jazz or twelve tone, the tipping point between whether I shell out the $1.99 to iTunes or shrug my shoulders and go on to the next piece of music is whether or not I find the music beautiful. When I apply this insight to the relationship between sequences like Fibonacci numbers and beautiful things we find in the natural world, I ask the question: Is it the math that defines an object as beautiful or is there a self-evident quality in nature, so powerful in its expression that we find hints of its inner beauty all around us, even in the human construct of mathematical equations?

I am much more at home with the idea that there is a fundamental, self-evident beauty, which happens to be consistent with human pursuits of truth like math and science. If we flip the narrative and think there is some type of "magic" to things like Fibonacci numbers, we run the risk of creating a "neo-Gnosticism" in which math and science are the only tools necessary to understand the world around us. But when we see a healthy balance and coexistence of science, art, music, literature, philosophy, and theology, we begin to understand that beauty points us toward a transcendent reality that is not conditioned by time and space: A transcendent reality we name Father, Son, and Holy Spirit.

As idealistic as these images of beauty may be, we need to acknowledge a great "distortion" amid our world of beautiful things. The Catholic tradition names this distortion Original Sin. In a fallen world, we must be careful not to presume that what is interpreted as pristine expressions of beauty are in and of themselves sufficient to explain the transcendent source of all beauty.

For example, one may feel that paintings done according to the tradition of Realism are the best expressions of truth because they communicate a clear, objective image that is self-evident to anyone who gazes upon them. However, the Impressionist would look at this same canvas and argue that this is just the "mask" hiding the inner, subjective truth of the scene that may look far different from the external appearance. The modern and post-modern artist may scoff at the very idea of beauty, presenting instead a canvas that is torn, stabbed, and splattered with paint, symbolically displaying a fundamental distrust in an objective source of goodness or beauty.

However, the Catholic could present this artist with the image of the crucified Christ, explaining how on this twisted, torn, pierced, and bleeding "canvas," we do not merely see an image of the death of beauty, but rather, as Hans Urs Von Balthasar put it, we also see "The Beautiful" himself, expressing his love which knows no bounds, even to the point of taking on the dark, twisted reality of our sinfulness and death.

Is beauty in the eye of the beholder? At one level, yes, beauty is in the eye of the beholder. However, in order to perceive beauty, there is an essential relationship in which the beauty of another touches the beauty within the beholder, creating a dynamic exchange that is named an experience of beauty. This exchange reminds me of St. Augustine's analogy of the Trinity as a mystical gaze of love between Father and Son in which the gaze "spirates" the Holy Spirit. In this sense, we can see a hint of God in every encounter of beauty, while acknowledging that sin can make this beauty distorted and more difficult to see.

Spiritual Exercise: As we contemplate the world around us, the world that is our planet and the world beyond our planet, we see elegant proportion and chaos, life and death, creation and destruction, which feeds the soul's yearning for beauty and meaning. How do you experience beauty in our natural world and what human expressions of this beauty draw you to its source?

THE DISCOVERY OF

GRAVITATIONAL WAVES

IS THERE A "MUSIC" TO THE UNIVERSE?

ONE OF THE GREATEST gifts God has blessed me with is a love of music. Classical music was my first field of study after high school, and I still have a deep affection for it and just about every other type of music from the ancients to the moderns. A serious study of music history reveals something far more interesting than Music Appreciation class. What the alert student is treated to is a fascinating weave of art, politics, philosophy, religion, and science. In a real way, music history is the study of the human experience through musical forms with special emphasis on God and creation, whether it be something as common as a flower or as extraordinary as the Resurrection.

I vaguely recall a lecture from one of my music history classes that explored Boethius' *De Institutione Musica*. The work presents the mathematical proportions we find in the sky above, on earth below, and in the human person, which became the theoretic structure of some of the greatest pieces of music ever written. This vision of the universe contributed to the idea of music emanating from the "Sacred Spheres" of creation, giving voice to a true harmony in the cosmos that also contained moral and ethical dimensions. The mathematical genius permeated every part of our culture from music, sculpture, architecture, literature, and just about any field of study that could be expressed mathematically. An idealistic vision of the interconnected nature of the world fueled an initial romanticism in me in college and made me feel that everything I encountered was singing a cosmic hymn of beauty and unity.

On February 11, 2016, the discovery of gravitational waves re-awakened for me the possibility of naturally occurring "music" in our universe. When two black holes merge, the event is so powerful that it actually creates a ripple in the fabric of space, sending out gravitational waves in all directions. This explanation may evoke the image of a pebble being dropped into a pool of water, sending out gentle ripples through the universe, but I am drawn to the "musical" aspect of these disturbances, intrigued by scientists who speak of "hearing a beat" in these waves. (I presume this is a different type of hearing than what we experience on earth because sound waves cannot be heard in space.)

Now, to be clear, if I were able to play the one-second pulses of what scientists of the LIGO team discovered, you would not mistake them for a Bach motet or a Beethoven symphony. If anything, it might remind you of an underwater event or something that sounds like a heartbeat through the stethoscope. As exciting as this litter "chirp" was, it may be just the tip of the gravitational iceberg because the LIGO team will be able to increase the sensitivity of its detector by about 27 to 30 times, giving hope that there is a lot more to these waves than initially discovered. The years ahead are very promising for gravitational waves. In the spirit of this discovery of "music" in the heavens, to quote Sonny and Cher, the beat goes on!

Why is this so significant? For that answer, I look to the many scientists who have been sharing this discovery with the world. Some have compared this to Apollo 11 and humanity landing on the moon. Others have equated it to Galileo and his first glimpse of our universe through his telescope. What history teaches us is that if this is on par with Apollo 11 and Galileo's telescope, we are on the precipice of new discoveries about the universe we live in. If, with time, this doesn't prove to be as significant as other moments in the history of science, we will at least have learned that Einstein's prediction of gravitational waves was shown to be true, solidifying this great scientist's place in history. In either case, we have exciting days ahead.

On a personal level, I have been reflecting on the question, What do I hope will come out of this discovery of gravitational waves? As a priest whose hobbies include music, art, and astronomy, my hope is that, just as in other eras of cultural history, this new way of understanding our world will contribute to new expressions of culture and our understanding of the world we live in. There has been a historic connection between science and the arts, as scientific advancements inspire musicians and artists to express these discoveries on canvas, in clay, and through music. Further, I hope these new cultural expressions will enrich our understanding of God's creation, giving us a better glimpse of the Creator. And, finally, I hope this will lead all of us to a humble disposition of heart, realizing we live in a world far more fascinating and marvelous than we can possibly imagine, pointing to a fascinating and marvelous God of whom we seek to understand, serve, and love.

PLUTO FLYBY

THE DWARF PLANET THAT DID NOT DISAPPOINT!

'M NOT A VERY good golfer. It may come as a scandal to some that a Catholic priest is not a scratch golfer, but the fact is, although I enjoy the walk and the fresh air, keeping a score card simply leads to depression. As a "weekend golfer," I enjoy the slightly uncharitable experience of watching PGA tournaments when the difficulty of the course promises scores that would look more like mine (if I kept score). Invariably, the winds end up being favorable to the players, the rains stay away, and a new course record is set, leaving some of us a little disappointed that we didn't see a colossal meltdown. However, there still is a reverential awe for professional golfers because, even if we took a sinful delight at their six shots from the sand trap, we know just enough about golf to understand that the "bad" shots they made would be impossible for the weekend golfer to even attempt.

I thought of professional golfers while watching a press conference about the New Horizons flyby of Pluto on July 14, 2015. Time and time again, I listened to professionals in astronomy answer questions from the news media with: "I don't know, we didn't expect this" and "We're going to have to wait in order to answer your question." As a hobby astronomer, there was a moment of "That's how I feel when I try to answer people's questions about astronomy." The interesting thing about the press conference, however, was that the scientists never looked discouraged, deflated, or defeated. Instead, their child-like excitement was so evident that, even though a lot of initial theories about Pluto and its moon Charon appear to have bitten the dust, there was joy about the opportunity to explore a "new world." Similar to the deference a weekend golfer gives to a PGA pro, I came away with tremendous respect for members of the science team, realizing that their I-don't-know-how-to-answer-your- question moments came from a profound understanding of the world we live in and were far beyond any "fairway shot" I could accomplish.

On March 17, 2016, NASA released a summary of the New Horizons team's discoveries about Pluto. As with many historic findings, some of the questions of initial intrigue were answered, while new questions emerged. And what were some of those findings? Pluto has been geologically active for about 4 billion years. It contains an ice plain larger than the state of Texas and that young (in geological terms). Pluto's moon Charon has an ancient surface that was likely formed through volcanic eruptions about 4 billion years ago and may point to an internal, frozen ocean that created the giant crack in Charon's crust. The surface of Pluto, rich with nitrogen, methane, and water, is far more complex than first thought, and the atmosphere is colder than originally suspected and contains nitrogen, methane, acetylene, and ethylene. There is much more that the New Horizons team found, but it's safe to say Pluto did not disappoint! (This list was taken from the NASA article: Top New Horizons Findings Reported in Science at http://www.nasa.gov/feature/top-new-horizons-findings-reported-in-science.)

Spiritual Exercise: When I look to apply this to faith, I'm reminded of a thought from Brother Guy Consolmango, director of the Vatican Observatory. He explains that when a scientist comes across a new discovery that questions a previously held idea, the scientist doesn't throw up his or her hands and say, "I don't believe in science anymore," but, instead, experiences the excitement of "Wow, there's something new here!" As people of faith, we can have our lives turned upside-down by experiences that make us question our faith, undoing what we initially thought was certain. Do we have the excitement of the scientist to see this as an opportunity to see God and the world in a new way or do we throw in the towel on our faith life out of frustration?

PIERRE GASSENDI

THE TRANSIT MERCURY, THE PRIEST WHO RECORDED THE DATA, AND EXPLORING WHERE WE GO FROM HERE

O N MAY 9, 2016, the world was able to enjoy the transit of Mercury. Planetary transits are quite rare because they only occur when the Sun, an inner planet (Mercury in this case), and the Earth are in alignment. The transit of Mercury occurs roughly 13 times each century, making this a noteworthy event. What is of particular interest from a Catholic perspective is that the first person to publish data pertaining to the transit of Mercury was a Catholic priest named Pierre Gassendi.

Gassendi lived from 1592 to 1655, overlapping the lives of Galileo and Kepler, and would have been in the autumn of his life at the time of Isaac Newton's birth (1643). He was known as a philosopher, astronomer, mathematician, and priest. His teaching career focused primarily on the philosophy of Aristotle, which he eventually rejected

.

in light of the emerging physics of his day. He also rejected Descartes' famous cogito ergo sum (I think, therefore I am). Gassendi's rejection of these two great thinkers was a move away from certainty in favor of reliability through sense experience. This led to Gassendi attempting to create a "Christian Materialism," trying to marry the Copernican understanding of the universe, the philosophy of Epicurus, and Catholic doctrine.

Central to Gassendi's attempt to achieve this complex weave of science, philosophy, and theology was atomism. The philosophy of atomism originated with Leucippus and was furthered by his disciple Democritus (born in the year 460 B.C.). Central to atomism was a materialist worldview that believed all things were made of atoms, which were eternal in nature, unable to be divided, infinite in number (which Gassendi rejected), and professed that there was no purpose or design to creation.

Gassendi's attempt to adapt this philosophy to Catholic Church teaching was founded on the idea that atoms had a God-given quality of self-motion (trying to address the "purposeless creation" belief of atomism). This argument rejected Aristotelian categories such as substance and accidents that also carried implications of meaning and purpose. Gassendi favored a view of the world that focused entirely on our sense experience of the universe but still defended the teaching authority of the Church, thus creating the foundation for what I am calling "Christian Materialism." Although there was much that Gassendi got right with science and theology, his historical significance is minimal with his ideas viewed as flawed by the scientific world, the philosophical world, and the Christian world. Nevertheless, his observation of the transit of Mercury, which confirmed Kepler's predictions of such an event, led to naming one of the craters on the moon after Gassendi.

I'm not an expert on Gassendi and cannot speak to the finer points of his life and thought. However, exploring the basics of Gassendi reminds me of the theme of the earlier chapter on Thomas Aquinas, wondering if the Church is in need of someone who can synthesize faith and science. As I observed the transit of Mercury, I remembered the man who attempted this very synthesis, trying to bring together empiricism and Catholic doctrine. What we find in Gassendi is a priest whose sincerity cannot be questioned, but his method can be.

A point of curiosity is how much influence Christian Materialism may have played in setting the backdrop for the "God of the Gaps" problem that emerged after Newton. If one tries to synthesize Christianity and materialism, there is a high risk that God becomes reduced to one being in creation, which the Church rejects, instead of God as "being itself" or "the pure act of 'to be.'" This reductionism of all things to the material can lead to the false notion that the only "space" for God's activity in the world is through the gaps of knowledge we have about how creation works. As our knowledge grows of the natural world, the gaps slowly shrink, forcing the "place" for God out of creation.

Ironically, I see a lot of this mentality being recreated today, especially in the intellectual spats between those of the Intelligent Design movement and the New Atheism. I, myself, fell into these traps when I first became interested in exploring the question faith and science, foolishly thinking that someday I would find a "God equation" that could explain everything, a Christian version of the Theory of Everything.

Over time, what I have learned is that God cannot and will not be found in an equation because God is not a part of mathematics but is that which allows our mathematical equations to exist in the first place. When understanding a figure like Pierre Gassendi, we can deeply admire his courage to marry two ideas that are so divergent. However, we can also learn from his errors, seeing in a marriage of Christian doctrine and empiricism a dangerous road of reducing all things to the material, including God.

Spiritual Exercise: What is your view of the relationship between God and creation? Do you try to make God one being in creation or do you allow God to be the grounding of your life? Ask the Lord in prayer to help you let go of the false images of God we can cling so tightly to so that God, in turn, can reveal Himself as not only being radically transcendent and mysterious, but also radically immanent, being ever present in our daily lives.

WHIRLPOOLS, SUNFLOWERS,

AND PINWHEELS

ONE OF MY FAVORITE things to do when I lived in Eau Claire, Wisconsin, was to stroll across the walking bridges that span the Chippewa and Eau Claire rivers. Originally built as train bridges, they are no longer in service. However, instead of tearing them down, the city, wisely, decided to repurpose them as pedestrian walking and biking bridges. These bridges and walking trails are so well done you feel like you have escaped the city, even though you are still in the middle of about 70,000 people.

When I was in a contemplative mood, I loved to walk to the middle of the bridge that stretches across the Chippewa River and watch the water flow beneath me. The Chippewa is quite shallow at this point and the water moves quickly. This shallow water, the current's speed, and a rocky river bed create the effect of hundreds of little "whirlpools" that spin down the river. It doesn't take much of an imagination to think of the galaxy that has been dubbed the "Whirlpool Galaxy," floating on the "river" of space-time, being twisted and contorted by forces we don't see or completely understand, but we know are present because of their impact on the visible universe. These moments become a spiritual experience, a sensation that God is allowing me to feel connected to something far beyond my comprehension (a galaxy) through a small spiral of water I can break apart with my fingers. These contemplative walks brought me great peace.

Another moment of connecting something small on earth with something unthinkably huge in space occurred while I was doing my classroom visit with the children at St. Joseph's Grade School, our parish grade school. I was sitting with our first-graders doing a "Q and A" session, which means that I would listen to them talk about their day and what they loved about life. One of the children, a young girl, raised her hand and said, "I love to look at the middle of sunflowers! They are so pretty, more pretty than anything else in the whole world!" Her innocent comment made me think of a "sunflower" I like to look at, the "Sunflower Galaxy." In this moment of excitement, I tried to put into first-grade language the visual similarities between the Sunflower Galaxy and the sunflower she so loves. As I saw confusion cloud her face, I asked, "Would it help if I bring a picture of the Sunflower Galaxy to class next time?" She quickly nodded her head up and down.

These serene images of whirlpools and sunflowers can evoke the predictable melancholy that comes at the end of summer for those of us who live in cold winter climates. The crisp, beautiful days of September are not only treasured for their beauty, but also their brevity. All too soon, the healing ambers and rusty yellows of fall, which paint our landscape with inescapably beauty, will depart. This brief time of serenity inspires many of us to go out and visit our local parks for one last deep breath of summer, reverting to simple joys like sitting on a blanket with family or friends for a picnic lunch. This pastoral image evokes another connection with the little things of life and the immensity of our universe, the simple play of a child holding a pinwheel in the wind. Can you imagine a "Pinwheel Galaxy" on a stick? I'm surprised the organizers of the State Fair haven't found a creative way to add this to their array of hand-held, deep-fried treats.

Why do people get drawn into astronomy? I'm sure there are as many answers to this question as there are people who have gazed into the heavens. I can't help but think that part of the answer to this question is that we can make easy, visual connections with what we see on earth and what we see in the heavens. Now, when exploring the science of these objects, there becomes a greater dissimilarity than similarity. Still, I wonder if the astronomer runs the risk desensitizing themselves to beauty in a way similar to what Mark Twain references when he writes of being a Mississippi riverboat captain, reducing a magnificent river into an endless litany of dangers.

No, the romance and the beauty were all gone from the river. All the value any feature of it had for me now was the amount of usefulness it could furnish toward compassing the safe piloting of a steamboat. Since those days, I have pitied doctors from my heart. What does the lovely flush in a beauty's cheek mean to a doctor but a "break" that ripples above some deadly disease? Are not all her visible charms sown thick with what are to him the signs and symbols of hidden decay? Does he ever see her beauty at all, or doesn't he simply view her professionally, and comment upon her unwholesome condition all to himself? And doesn't he sometimes wonder whether he has gained most or lost most by learning his trade? ("Two Ways of Seeing a River" by Mark Twain)

I can see an important application for both faith and science: If our theological and scientific pursuits become so divorced from the common experience of everyday people, creating a dense web of theories and ideas that are close to incomprehensible even to the brightest of minds, does this really contribute to the human condition? Am I arguing that we shouldn't have advanced science and theology? No, we need to continually explore our understanding of the physical world and how God inspires us to understand the world we live in. However, even the most complicated aspects of faith and science need to be made accessible to all people, allowing for public discussion that includes science, theology, philosophy, the arts, literature, and all branches of knowledge.

Put another way, we all should have the heart of the first-grader, willing to experience the intricate beauty of our world, as she does when looking at sunflowers, and share that beauty with others. For those who are in the professional sciences and theology, it's our responsibility to invite this young mind to comprehend a world that is both connected with the beauty she sees and deepens her knowledge in new ways, never forgetting the initial wonder that drew her (and us) into this exploration in the first place. In the process, we are called to walk together as a community, constantly seeking to understand the beauty, goodness, and truth of the world we live in.

Spiritual Exercise: What connections do you see between the simple things of our world and the universe that exists beyond our common home? Ask God to reveal His beauty to you in creation, allowing you a moment of awe at simple beauty that also contains within its aesthetics a complexity of truth, calling the human mind to study and understand our world in greater detail.

CONCLUSION

AND THE PILGRIMAGE CONTINUES!

As we come to the end of this collection of reflections on faith, science, and creation, a logical question is: Where do we go from here?

Truly, the exploration of faith and science never really ends. There will always be advancements in our scientific understanding of the world, and God will continue to inspire the faithful to seek a deeper understanding of our faith and apply that understanding to the world around us.

The important question for people of faith and people of science is: Do we pursue the truths of our world as dialogue partners or do we continue to find ways to separate ourselves from one another?

In my priestly ministry, I have been blessed to know many faithful Catholics of science who are not only open to this dialogue but sincerely thankful when it takes place. Yes, I encounter the occasional moments of tension, like when students in my Newman ministries share experiences of gross misstatements made about the Church in some of their science classes that can only be categorized as dishonest and hateful. Nevertheless, I find those moments to be few and far between. What I find more often are people who want to embrace both faith and science on their own terms, but they simply are afraid to because the "popular narrative" of faith and science as enemies has become so engrained in their psyches.

That being said, we, too, as Catholics, need to realize there are ugly events in history we must confront if a healthy relationship between faith and science is to be achieved. However, let us not allow these historical difficulties overshadow the rich history of how faith and science have walked with one another as dialogue partners.

We need to celebrate the example of Pope Benedict XIV who, at the height of the Enlightenment, encouraged not only the advancement of the emerging science of the day, but helped break historic ground by encouraging a young woman by the name of Laura Bassi to pursue the natural sciences. As we read earlier, Bassi eventually became the first woman to teach at the University of Bologna and was appointed to Benedict's scientific elite called the Benedettini.

We should celebrate priests like Fr. Pierre Gassendi, the first to record the transit of Mercury, and modern scientist Fr. Stanley Jaki who rightly questioned the ability of physics to achieve a "Theory of Everything." And we should celebrate Monsignor Georges Lemaitre who had the courage to see in Einstein's equations something far different than the static universe affirmed by most scientists of that time. Lemaitre saw a universe that was expanding, which meant that at one time it was infinitely small. Although his ideas were mockingly called "The Big Bang," Lemaitre's theory is still the foundation of our modern understanding of the universe.

There are more examples, but the main point is that the Church has continued to embrace science on science's own terms. This does not mean, of course, that this mutual pursuit has not had moments of apparent disagreement and contradiction. But it's important to remember that the Church believes that these apparent contradictions are just that, apparent, since, in the final analysis, truth cannot contradict truth.

Freed from the misconceptions of the past and optimistic about our future, we should encourage, as a Church, a renewed fraternal charity between faith and science, encouraging people of faith and people of science to walk together on a pilgrimage of truth.

On a personal level, I have seen my exploration of faith and science as a pilgrimage as I consider the billions of light years that have provided mystery and intrigue. In the process, I have come to learn that the proper disposition of heart on this journey is to humble ourselves between the two great mysteries of God and the universe, realizing our inability to fully grasp either. On this journey, we begin to realize that what we learn the most when exploring faith and science is something elusive within ourselves, intimately connected to how we see God and the world we live in: Our life's meaning and purpose.

A pilgrimage can be harmed by not being open to the journey. If someone thinks he or she has God and the world figured out, what's the point of any further exploration? When humility before that which is beyond comprehension is replaced with arrogance, the human heart can become hardened, dismissing such a pilgrimage of faith and science as a "waste of time."

As a priest, many people confide in me that as they grow older the lack of exploring such questions begins to gnaw at them. There is regret that they chose to make life more about the practical things instead of pursuing questions of meaning, purpose, beginnings, and endings.

An interesting manifestation of this internal angst is the many online and DVD continuing education courses one can find that are designed for adults. What courses are most commonly found? Philosophy, theology, astronomy (or some other science), and a myriad of titles trying to help people find meaning in life.

If we don't allow ourselves the experience of a meaningful,

pilgrimage-type of exploration in our lives, eventually, the pilgrimage seeks us out as the human heart aches for meaning and purpose. A pilgrimage is not only a "nice idea" but something essential to the human experience.

One of my favorite quotes from the Prayer Journal of Flannery O'Connor is this, "No one can be an atheist who does not know all things. Only God is an atheist. The devil is the greatest believer and he has his reasons."

One of the greatest gifts I receive when I pray or gaze into the heavens is the humble recognition that my knowledge of faith and science is, to borrow a sentiment from St. Thomas Aquinas, nothing more than straw in the eyes of God. Nevertheless, God calls us to pilgrimage, to understand the world we live in and our relationship with Him who is our Source and Summit.

I want to thank you for embracing this pilgrimage in this book of reflections and invite you to continue the journey. Together, may we delve deeper into the never-ending adventure of discovering who God is and our place in the universe, seeking to find meaning, purpose, and peace while being eternally drawn into the love of God.

REFERENCED TEXTS AND MEDIA

All Biblical quotations are taken from the online, New American Bible translation provided by the United States Council of Catholic Bishops (USCCB) and cited in the body of the text http://www.usccb.org/bible/books-of-the-bible/index.cfm

All Catechism Citations are taken from the online, Catechism of the Catholic Church (CCC) and cited in the body of the text. (Vatican City: Libreria Editrice Vaticano 1993) http://www.vatican.va/archive/ENG0015/_INDEX.HTM

Chapter One

St. John Paul II, *Fides Et Ratio*. (Vatican City: Libreria Editrice Vaticana 1998) Opening Blessing http://w2.vatican.va/content/john-paul-ii/en/encyclicals/documents/hf_jp-ii_enc_14091998_fides-et-ratio.html
NSTA Position Statement: The Nature of Science, National Science Teachers Association Website. http://www.nsta.org/about/positions/natureofscience.aspx
Teaching about evolution and the nature of science. Online Open Book. (National Academies Press 1998) http://www.nap.edu/read/5787/chapter/6#58

Chapter Two

C.S. Lewis, *Reflections on the Psalms*. (New York: Harcourt, Brace & Company 1986)
The Collegeville Bible Commentary: Old Testament. (Collegeville: The Liturgical Press 1986)
The Jerusalem Bible. Footnotes Genesis 1:16 (New York: Doubleday 1966)

The Catholic Study Bible: New American Bible. (New York: Oxford University Press 1990)

Pope Emeritus Benedict XVI, *'In the beginning...' A Catholic Understanding of the Story of Creation and the Fall.* (Grand Rapids: William B. Eerdmans Publishing Company 1990)

Chapter Three

William E. Carroll, "Creation, Evolution, and Thomas Aquinas." *Revue des Questions Scientifiques* 171 (4) 2000: 319-347 Reprinted with author's permission at http://www3.nd.edu/~afreddos/ courses/43150/carroll3.htm

Stephen Hawking, *A Brief History of Time.* (New York: Bantam Press 198)

Stephen Hawking; Leonard Mlodinow, *The Grand Design.* (New York: Bantam Books 2010)

Barr, Stephen M., "Much ado about 'nothing': Stephen Hawking and the Self-Creating Universe," *First Things.* Web Exclusives (September 10, 2010) http://www.firstthings.com/web-exclusives/2010/09/much-ado-about-ldquonothingrdquo-stephen-hawking-and-the-self-creating-universe

Chapter Four

"Creation," *The New Dictionary of Theology.* Editors Komonchak, Joseph; Collins, OSB, Mary; Lane, Dermot, (Collegeville: Liturgical Press 1987).

Chapter Five

Christoph Schonborn, "Finding Design in Nature," *The New York Times.* July 7, 2005 http://www.nytimes.com/2005/07/07/opinion/finding-design-in-nature.html?_r=1

Stephen M. Barr, "The Design of Evolution," *First Things.* October

2005 http://www.firstthings.com/article/2005/10/the-design-of-evolution

"Communion and Stewardship: Human Persons Created in the Image of God," *International Theological Commission*. http://www.vatican.va/roman_curia/congregations/cfaith/cti_documents/rc_con_cfaith_doc_20040723_communion-stewardship_en.html

Chapter Six

Irenaeus of Lyons, *The Demonstration of the Apostolic Preaching*. http://www.ccel.org/ccel/irenaeus/demonstr.iv.html

Irenaeus of Lyons, Against Heresies. Book 4 http://www.earlychristianwritings.com/text/irenaeus-book4.html

Irenaeus quote, "The glory of God is man fully alive, and the life of man is the vision of God!," take from: Delhaye, Philippe, "Pope John Paul on the contemporary importance of St. Irenaeus," *L'Osservatore Romano*. Weekly English Edition. February 9, 1987. Reprinted with permission at: https://www.ewtn.com/library/Theology/IRENAEUS.HTM

Chapter Seven

Mark Midbon, "'A Day Without Yesterday', Georges Lemaitre & the Big Bang." Commonweal Magazine Vol. 127 No. 6 (March 24, 2000) 18-19. Reprinted with authors permission by Catholic Education Resource Center. http://www.catholiceducation.org/en/science/faith-and-science/a-day-without-yesterday-georges-lemaitre-amp-the-big-bang.html

"Sir Fred Hoyle; Coined 'Big Bang,'" Obituaries, from the Times Staff and Wire Reports, Los Angeles Times. (August 23, 2001) Online version: http://articles.latimes.com/2001/aug/23/local/me-37483

Joseph J. Laracy, "The Faith and Reason of Father George Lemaitre," Homiletic & Pastoral Review. Ignatius Press Reprinted

online with permission by CatholicCulture.org https://www.
catholicculture.org/culture/library/view.cfm?recnum=8847

Pope Pius XII, "The Proofs from the Existence of God in the Light
of Modern Natural Science." Address of Pope Pius XII to the
Pontifical Academy of Sciences. November 22, 1951. Reprinted
by Papal Encyclicals Online http://www.papalencyclicals.net/
Pius12/P12EXIST.HTM

"ATV-5: Georges Lemaitre, Monseigneur Big Bang," European
Space Agency (ESA) July 28, 2014 Youtube video: https://www.
youtube.com/watch?v=RL6ndOAOEeE

Chapter Eight

Stanley Jaki, "The Mind and Its Now," Metanexus Institute https://
vimeo.com/10588094

Stanley Jaki, "On a discovery about Godel's Incompleteness Theo-
rum," Pontifical Academy of Sciences, Acta 18, Vatican City
2006. http://www.sljaki.com/2006-03-incompleteness.pdf

Steven Hawking, "Godel and the End of the Universe," Hawking.
org http://www.hawking.org.uk/godel-and-the-end-of-physics.
html

Chapter Nine

Megan Geraghty (Lobos), "Making Multicultural Relationships
Work: Our Experience as a Chilean-American Couple." For
Your Marriage, An initiative of the United States Conference
of Catholic Bishops. http://www.foryourmarriage.org/making-
multicultural-relationships-work-our-experience-as-a-chilean-
american-couple/

St. John Paul II, "Women: Teachers of Peace," Message of His
Holiness Pope John Paul II for the XXVIII World Day of
Peace. Copyright Librera Editrice Vaticana http://w2.vatican.
va/content/john-paul-ii/en/messages/peace/documents/

hf_jp-ii_mes_08121994_xxviii-world-day-for-peace.html

Voices of Faith Website. http://voicesoffaith.org

Erin Blakemore, "These little-known nuns helped map the stars," Smithsonian.com http://www.smithsonianmag.com/smart-news/ these-little-known-nuns-helped-map-stars-180959012/?no-ist

Pope Emeritus Benedict XVI, Apostolic Letter proclaiming Hildegard of Bingen, professed Nun of the Order of Saint Benedict, a Doctor of the Universal Church. Copyright Librera Editrice Vaticana https://w2.vatican.va/content/benedict-xvi/en/ apost_letters/documents/hf_ben-xvi_apl_20121007_ildegarda-bingen.html

Women in Science. Directorate – General for Research (Luxemburg: Office for Official Publications of the European Communities 2009) https://ec.europa.eu/research/audio/women-in-science/pdf/wis_en.pdf

Dan Stober, "Papers of Europe's first female professor to become available online, with help from Stanford Libraries," Stanford News Online. http://news.stanford.edu/news/2012/january/ libraries-scan-bassi-010412.html

Chapter Ten

C.S. Lewis and Pauline Baynes, *The Magicians Nephew*. (New York: HarperCollins 2000)

Catherine Pickstock, *After Writing: On the Liturgical Consummation of Philosophy*. (Oxford: Blackwell Publishers 1998)

Homily of His Holiness Benedict XVI, Celebration of Vespers with the Faithful of Aosta (Italy). Cathedral of Aosta, Friday July 24, 2009. https://w2.vatican.va/content/benedict-xvi/en/homilies/2009/documents/hf_ben-xvi_hom_20090724_vespri-aosta. html

Chapter Eleven

David Fagerberg, "Humility Without Humiliation: A Capacitation for Life in Elfland in the Thought of G.K. Chesterton," American Theological Inquiry: A Biannual Journal of Theology, Culutre, and History. (Minneapolis: Volume 6, No. 1 2013) 9-21 Youtube Lecture available at: https://www.youtube.com/watch?v=e721eECFlCg

G.K. Cheterston, *Tremendous Trifles*. http://www.gutenberg.org/ebooks/8092?msg=welcome_stranger

Brenda Frye, "Looking back in time..." The Catholic Astronomer Blog. http://www.vofoundation.org/blog/looking-back-in-time/

Chapter Twelve

Flannery O'Connor, "The Church and the Fiction Writer," America Magazine. (March 30, 1957) Online http://americamagazine.org/issue/100/church-and-fiction-writer

Chapter Thirteen

Carl Sagan, "Pale Blue Dot," Cosmos. (Copyright 1980 by Carl Sagan Productions, Inc.) Excerpt found online at: http://www.planetary.org/explore/space-topics/earth/pale-blue-dot.html

Chapter Fourteen

Tim Noonan and R.E. Houser, "Saint Bonaventure," Stanford Encyclopedia of Philosophy. (Stanford: The Metaphysics Research Lab 2014) Online http://plato.stanford.edu/entries/bonaventure/

Pope Francis, Laudato Si'. (Vatican City: Libreria Editrice Vaticana 2015) Online http://w2.vatican.va/content/francesco/en/encyclicals/documents/papa-francesco_20150524_enciclica-laudato-si.html

Father James, Salmon, SJ, Ph.D. *The Intersection of Science and Theology: Evolutionary Theory and Creation.* (NYKM 2014) Disc 4, Lecture 12

Dr. Michio Kaku, "Is God a Mathematician?" Big Think (Published January 2013 Online: https://www.youtube.com/watch?v=jremlZvNDuk)

Dr. Michio Kaku, "Michio Kaku's opinion on God." Published Online 2011 https://www.youtube.com/watch?v=SBB2qHgZvLY

Chapter Fifteen

Anton G. Pegis (editor), Basic Writing of St. Thomas Aquinas: Vol. 1. God and the Order of Creation. (Cambridge: Hackett Publishing Company 1997)

Thomas Aquinas, Summa Contra Gentiles.

Bishop Robert Barron, Bishop Barron on Thomas Aquinas and the Argument from Motion. (Word on Fire Ministries: Youtube) https://www.youtube.com/watch?v=bdjjqFSEJ_Y

Chapter Seventeen

Rev. William Suaders, "The History of the Advent Wreath," Arlington Catholic Herald. Reprinted with permission at Catholic Education Resource Center: http://www.catholiceducation.org/en/culture/catholic-contributions/the-history-of-the-advent-wreath.html

Maximus the Confessor, Selected Writings: Classics of Western Spirituality. John Forina (editor). (New Jersey: Paulist Press 1985)

Hans Urs Von Balthasar, Cosmic Liturgy: The Universe According to Maximus the Confessor. (San Francisco: Saint Ignatius Press 2003)

Most Rev. Peter J. Elliot, "The Glory of the Liturgy: Pope Benedict's Vision," Keynote Address at Time Drawn Into Eternity: Sacred Time and the Liturgical Calendar. Reprinted online by Catholic

News Agency http://www.catholicnewsagency.com/resources/ roman-missal-3rd-edition/bishops/the-glory-of-the-liturgy-pope-benedicts-vision/

Andrew McGowan, "How December 25 Became Christmas," Bible History Daily: Biblical Archeology Society. (Published Online December 2, 2015) http://www.biblicalarchaeology.org/daily/biblical-topics/new-testament/how-december-25-became-christmas/

Joseph Cardinal Ratzinger, The Spirit of the Liturgy. (San Francisco: Ignatius Press 2000)

Chapter Nineteen

Brother Guy Consolmagno, Was There Really A Star. (Catholic News Service: Youtube Video) https://www.youtube.com/ watch?v=WsQgRMokAek

Bishop Robert Barron, Bishop Barron on Christmas. Word of Fire Youtube: https://www.youtube.com/watch?v=eUW-A2qZzWY

Chapter Twenty-one

Exsultet: The Proclamation of Easter. (Roman Missal: United States Council of Catholic Bishops) Online: http://www.usccb.org/ prayer-and-worship/liturgical-year/easter/easter-proclamation-exsultet.cfm

Martin Luther King Jr., Where Do We Go From Here? Chaos or Community. (Boston: Beacon Press 1986)

Chapter Twenty-six

"Explosive Planet Io," Wonders of the Solar System. (BBC2: Youtube March 24, 2010) https://www.youtube.com/ watch?v=rAejrPirGxI

"Science Cast: Close Encounter with Enceladus," (Science at NASA: Youtube October 27, 2015) https://www.youtube.com/watch?v=nts-bkhoMt4

"Europa: Ocean World," (NASA Jet Propulsion Laboratory: Youtube November 21, 2014) https://www.youtube.com/watch?v=kz9VhCQbPAk

Chapter Twenty-eight

Press Conference, "LIGO detects gravitational waves (part one)," (National Science Foundation: Youtube February 11, 2016) https://www.youtube.com/watch?v=aEPIwEJmZyE

Chapter Twenty-nine

Press Conference, "NASA News Conference on the New Horizons Mission," (NASA: Youtube July 17, 2015) https://www.youtube.com/watch?v=xAGwxl7FZWw

Peregrino Press (that is, "Pilgrim" Press translated from Spanish to English) is a media publishing company in De Pere, WI. Born out of a desire to share our stories and our faith more intimately, Peregrino Press publishes engaging and inspired con-tent that enhances and promotes our Catholic identity.

As children of God – and as brothers and sisters in Christ – we all journey together. The image of being on a pilgrimage finds meaning in walking with others on life's journey, for no one of us is meant to "go-it-alone." Simply, in Christ, we are pilgrims and friends on a profound journey of faith. And it is in the sharing of the stories of our lives that we come to new life, hope, and peace.

To Heaven & Back

The Journey of a
Roman Catholic Priest

While connecting me up to various machines and monitors and placing nitroglycerine tablets under my tongue, the doctor looked at me early on and said: "You waited too long to get here, sir." I was petrified. There was no immediate response on my part, only a deep sense of fear en-tombed by a completely paralyzed inner spirit. Much to my surprise, he spoke again: "Sir, you've waited too long to get here. You're not going to make it." And he continued: "If you believe in God, this is the time to make peace." In that very moment, I said to myself, "What do you mean, 'If you believe in God?' Of course I believe in God! I'm a man of faith and a Catholic priest!"

Fr. John Tourangeau, O. Praem., a Norbertine priest who had an afterlife experience following a major heart attack, emphatically states, "Heaven is for real!" Within this enlightening and hope-filled book, Fr. John weaves a powerful and dynamic tapestry of the Kingdom of God at hand through the exploration of Christian tradition, Sacred Scripture, Catholic teaching, as well as his own lived experiences. "While the full-ness of heaven cannot be fully experienced in our life here on earth," Father explains, "we are able to more fully experience God's love for us through our relationships with others. For it is in and through these rela-tionships that we draw closer to Christ and his promise for us."